The
Mayan
Calendar
and the
Transformation
of Consciousness

Kukulcan, the Plumed Serpent, the Mesoamerican deity called
Quetzalcoatl by the Aztecs. He symbolizes light as well as duality
and is the carrier of the energy 9 Ik in the Sacred Calendar.
Kukulcan played a major role in the worldview of the ancient
civilizations of Mexico and came in many guises, such as the wind
god depicted here. He also incarnated in human form at times,
and because of his nature as the principle of light, it has been
suggested that his earthly incarnations were manifestations of the
same energy as that of Christ.

The
Mayan
Calendar
and the
Transformation
of
Consciousness

Carl Johan Calleman, Ph.D.

Foreword by José Argüelles

Bear & Company
Rochester, Vermont

Bear & Company
One Park Street
Rochester, Vermont 05767
www.InnerTraditions.com

Bear & Company is a division of Inner Traditions International

LIBRARY OF CONGRESS CATALOGING-IN-PUBLICATION DATA

Calleman, Carl Johan.
  The Mayan calendar and the transformation of consciousness / Carl Johan Calleman.
     p. cm.
  Includes bibliographical references and index.
  ISBN 1-59143-028-3 (pbk.)
  1. Maya calendar. 2. Maya chronology. 3. Maya cosmology. I. Title.

F1435.3.C14C35 2004
529'.32978427—dc22

2004003427

Printed and bound in the United States by Lake Book Manufacturing, Inc.

10 9 8 7 6 5

Text design by Priscilla Baker
Text layout by Virginia Scott Bowman
This book was typeset in Apolline, Avenir, and Kabel with Antique Olive and Kabel as the display typefaces

Illustration credits:
Frontispiece, fig. 3.9, fig. 5.7, Codex Vindobonensis, Mixtec; fig. 1.3, Dresden Codex, postclassic Mayan, Yucatán; fig. 1.5, fig. 3.1, fig. 3.14, courtesy of Schele and Freidel, A Forest of Kings; fig. 1.6, fig. 2.8, fig. 4.3, photos by the author; fig. 2.3, Codex Borbonicus, Aztec, sixteenth century; fig. 2.4, courtesy of the Egyptian Tourist Board; fig. 3.12, courtesy of Bob G. Lind; fig. 5.1, courtesy of Freidel, Schele, and Parker, Maya Cosmos; fig. 5.8, Codex Vaticanus A; fig. 5.14, Codex Telleriano-Remensis, Aztec, sixteenth century; fig. 9.4, "Early Transactions of the American Philosophical Society," The American Magazine, Philadelphia, 1769; fig. B.1, Codex Magliabechiano, Aztec, sixteenth century; fig. B.4, Codex Borgia, late postclassical period; figs. 1.8–1.10, 2.6, 3.2–3.8, 3.11, 4.5, 4.6, 5.2, 5.13, 6.1, 6.4–6.7, 7.1, 7.3, 7.4, 9.1–9.3, 9.6, A.2, A.3, B.7, D.1, courtesy of Ola Petterson

To Quetzalcoatl, the heavenly Creator
of the Sacred Calendar,
and to the people of the Maya,
carriers of its truth.

Contents

Foreword

Since the Harmonic Convergence of August 16–17, 1987, and the concurrent publication of my book, *The Mayan Factor: Path Beyond Technology,* interest in Mayan civilization and, in particular, the Mayan calendar has become widespread around the planet. Not only have there since appeared numerous books on the Mayan prophecies and civilization in general, but there has been a revival of the traditional Mayan time-knowledge as well, most notably through the efforts of Alejandro Oxlaj (Cerillo) of the Quiché-Maya in Guatemala, and, in Mexico, of Hunbatz Men of the Yucatec Maya. A deeper reason for this interest is that the end of the Mayan Great Cycle of thirteen *baktuns* will soon draw to a close on the winter solstice (Northern Hemisphere) 2012. It is awareness of this conclusive date that, acting like a signal in the human DNA, prompts so much interest and enthusiasm in the Mayan calendar, as is demonstrated, for instance, by John Major Jenkins's *Maya Cosmogenesis 2012.*

Carl Johan Calleman's *The Mayan Calendar and the Transformation of Consciousness* follows in the tradition of *The Mayan Factor* in being an in-depth philosophical investigation and interpretation of the Mayan calendar, showing its absolute relevance as a tool and guide for this final stage of the thirteen-baktun Great Cycle. As Calleman well demonstrates, the Mayan calendar is a multivalued system encompassing a deep philosophy of nature and natural rhythms, which in turn afford applications for historical analysis. Calleman is acutely aware of the religious and theological ramifications of his interpretations.

Not the least of the ramifications stems from the infamous Christian book-burning of the Mayan texts in 1562. As Calleman writes in chapter 4,

> The higher perspective generated by the Mayan calendar . . . cannot be turned into a new dogma. . . . Rather, the Mayan calendar provides a possible framework for the common exploration by individuals who share a respect for the contributions and views of others. The Mayan calendar, properly understood, is in its essence alien to all fundamentalism, to everyone believing that there is but one true religion that holds the whole truth. Sadly, this is why the Catholic priests burned all the books of the Maya: They threatened the monopoly on the truth that Catholicism at the time was striving to establish. Today, however, increasing numbers of people are turning away from organized religion out of a desire to seek the truth for themselves. This facilitates the revival of the Mayan calendar on a worldwide scale among all those aspiring to the unity of humanity.

This point of view is broadly ecumenical and universal and speaks of the most spiritual nature of the Mayan calendar. It also speaks of Calleman's fascinating effort to create a grand synthesis of humanity's spiritual history—a history of the human mind—and an analysis of the present and near future that is based on a strict adherence to the Long or True Count, including the value of the 360-day *tun* cycles. Equally fascinating is Calleman's analysis and interpretation of the thirteen baktuns as alternating day-and-night cycles, and as related to the cosmology of the Thirteen Heavens. In this regard he sees that we are passing through a Galactic Underworld, and in 2011 we enter into a Universal Underworld.

But for Calleman, as we approach the 2012 date, what is most important is that he sees a great spiritual awakening and unification occurring. "Based on our knowledge of the Mayan calendar," he writes in chapter 9, "there is little doubt that as we approach the completion of the cosmic plan, the number of enlightened people will increase greatly. The proof from the Mayan calendar that the mind has a history offers great hope to humankind."

What I find interesting about Calleman's work is that it substantiates the premise of *The Mayan Factor,* that the Mayan civilization and the calendar in particular are the overlooked factors in any consideration of the course and history of human civilization. Calleman has been able to explore and investigate from his perspective the depths of the Mayan calendar, demonstrating its fundamental spiritual and mental nature, which speaks to a level of synthesis that the world sorely needs today. Even more important, Calleman sees the Mayan calendar in relation to the divine or cosmic plan. "The Mayan calendar . . . ," he writes in chapter 9, "remains our most important instrument for studying the cosmic plan."

It is also interesting that the Mayan calendar inspires in Calleman a genuinely holistic and global understanding of ourselves. For though the work I have pursued in my investigation of the Mayan calendar differs in certain points from that of Calleman's—and it is wise to be open and fearlessly study all points of view—we have in common methods of applying fractal units of the calendar as tools of analysis. For both of us it is clear that the Mayan calendar is well disposed to fractal holographic applications based on the key numbers—9, 13, and 7. The reader will find Calleman's fractal-based interpretations, utilizing many graphics, to be thought-provoking and stimulating.

Finally, I would like to emphasize Calleman's perception of the two critical Venus passages that will soon be occurring. The first will occur on the date June 8, 2004, and the second on June 6, 2012. These dates, marking the passage of Venus in front of the sun, are for Calleman an augur of the "final transformation of the human mind into a mind of light." Venus passages always occur in pairs. The last two pairs occurred in 1874/1882, and before that 1761/1769. Of course, the dates 2004 and 2012 are most significant. The 2004 date must be seen as the herald of the Great Calendar Change, July 25–26, 2004, while the second passage in 2012 initiates the closing of the cycle. For the 2004 date, Calleman is calling for a worldwide meditation. This meditation would be a great opportunity to set the stage for the spiritual triumph of the Great Calendar Change. While on the True Count June 8, 2004, is 6 Ehecatl (wind), a sign of Quetzalcoatl, on the Dreamspell count that day is 3 Monkey, one of the Thirteen Clear Signs on the tomb of Pacal Votan. And it will occur but one week before the fifty-second anniversary of the opening of Pacal Votan's tomb.

Pondering the meanings of Calleman's book, let us all search our hearts and minds and seek the higher unification of the spirit that is afforded by the study and practice of the Mayan calendar in all its forms. Let us prepare to use the Mayan calendar as a tool to study the course of events generated by the cosmic plan, for are we not in the end all fashioned of one soul? Calleman's text on the Mayan calendar can only bring whoever reads it to a new threshold of understanding. Let us all move forward as one to the One!

José Argüelles

Valum Votan, Closer of the Cycle
author of *The Mayan Factor* and *Time and the Technosphere*
Overtone Moon 28, Kin 186, Yellow Rhythmic Star
True Count: Uaxac Etznab
Gregorian: December 12, 2002

Acknowledgments

I want to thank Ola Petterson for critically reading the manuscript and making valuable suggestions to make my ideas visible.

	National Underworld	Planetary Underworld	Galactic Underworld	Universal Underworld
Ruling Energy	13 *baktun* 5,125 years 13 Days/Nights of 394.3 years	13 *katun* 256 years 13 Days/Nights of 19.7 years	13 *tun* 12.8 years 13 Days/Nights of 360 days	13 *uinal* 260 days 13 Days/Nights of 20 days
Day 1 is Heaven 1 **Sowing** *Xiuhtecuhtli,* god of fire and time	Aug. 11, 3115 2721 BCE	July 24, 1755– 1775	Jan. 5, 1999– Dec. 31, 1999	Feb. 11, 2011– March 3
Night 1 is Heaven 2 **Inner Assimilation of New Wave** *Tlaltecuhtli,* god of earth	2721–2326	1775–1794	Dec. 31, 1999– Dec. 25, 2000	March 3– March 23
Day 2 is Heaven 3 **Germination** *Chalchiuhtlicue,* goddess of water	2326–1932	1794–1814	Dec. 25, 2000– Dec. 20, 2001	March 23– April 12
Night 2 is Heaven 4 **Resistance against New Wave** *Tonatiuh,* god of the sun and the warriors	1932–1538	1814–1834	Dec. 20, 2001– Dec. 15, 2002	April 12– May 2
Day 3 is Heaven 5 **Sprouting** *Tlacoteotl,* goddess of love and childbirth	1538–1144	1834–1854	Dec. 15, 2002– Dec. 10, 2003	May 2– May 22
Night 3 is Heaven 6 **Assimilation of New Wave** *Mictlantechutli,* god of death	1144–749	1854–1873	Dec. 10, 2003– Dec. 4, 2004	May 22– June 11
Day 4 is Heaven 7 **Proliferation** *Cinteotl,* god of maize and sustenance	749–355	1873–1893	Dec. 4, 2004– Nov. 29, 2005	June 11–July 1
Night 4 is Heaven 8 **Expansion of New Wave** *Tlaloc,* god of rain and war	355–40 CE	1893–1913	Nov. 29, 2005– Nov. 24, 2006	July 1–July 21
Day 5 is Heaven 9 **Budding** *Quetzalcoatl,* god of light	40–434	1913–1932	Nov. 24, 2006– Nov. 19, 2007	July 21– Aug. 10
Night 5 is Heaven 10 **Destruction** *Tezcatlipoca,* god of darkness	434–829	1932–1952	Nov. 19, 2007– Nov. 13, 2008	Aug. 10– Aug. 30
Day 6 is Heaven 11 **Flowering** *Yohualticitl,* goddess of birth	829–1223	1952–1972	Nov. 13, 2008– Nov. 8, 2009	Aug. 30– Sept. 19
Night 6 is Heaven 12 **Fine tuning of New Protoform** *Tlahuizcalpantecuhtli,* god before dawn	1223–1617	1972–1992	Nov. 8, 2009– Nov. 3, 2010	Sept. 19– Oct. 9
Day 7 is Heaven 13 **Fruition** *Ometeotl/Omecinatl,* Dual-Creator God	1617– Oct. 28, 2011	1992– Oct. 28, 2011	Nov. 3, 2010– Oct. 28, 2011	Oct. 9– Oct. 28, 2011

The Calleman Matrix, a prophecy chart showing the periods of rule by the thirteen deities (energies) in the four highest Underworlds. This chart may be used for encyclopedic studies of the evolution of all aspects of human history. It also provides a basic matrix for prophecy.

Preface

The Mayan calendar shares many of its messages with other spiritual traditions: "We are all One," "Life has a purpose," and "God is Love." Yet it should not be overlooked that it also conveys a unique message: There is a deadline for the creation of the enlightened golden age at "the end of time," and we all need to participate as co-creators in that process. This is the crucial message of the Mayan calendar that needs to be assimilated today. Many distortions and misunderstandings, as well as a common skepticism toward prophecy in general, have, at least until now, blocked many from fully recognizing the divine time plan.

One typical misunderstanding is that the Mayan calendar is of interest only to those interested in Mayan culture. Of course, the ancient Maya were the first people to develop and study the calendar described in this book and should be given credit for that. Yet today interest in anthropology or ancient cultures is hardly a very compelling reason to study the Mayan calendar, a prophetic source that concerns every member of our species. The Mayan calendar carries a universal truth that is probably a much more urgent topic for study in today's global community than it ever was to the Maya. The present book, *The Mayan Calendar and the Transformation of Consciousness,* is a result of my work to extract this universal truth from the Mayan calendar system and present it in a way that is meaningful to modern people. In the process I hope to show that this calendar also illuminates the many other religious and philosophical traditions that have emerged on our planet.

The Mayan calendar is a prophetic tradition. Prophesy, or the ability to make predictions about the future, is an art that has been looked upon very differently in different quarters and at different times. Astrology, kabbalah, numerology, and tarot are examples of divinatory tools that have been—and still are—used all over the world by ordinary people as well as rulers. The readings of Nostradamus and Edgar Cayce, the Book of Revelation, the cryptograms described in *The Bible Code,* and the Hopi Prophecy have been proposed to be prophetic in content, and countless interpretations have purported to reveal their hidden messages. Most of these sources cannot be said to be unambiguously true. Their lack of a timeline makes it impossible to judge whether the predictions made in them are really to the point or are merely the afterthoughts of an interpreter.

The timeline of the Mayan calendar, by contrast, is unambiguously true and not of a hocus-pocus nature. Anyone with access to a standard encyclopedia can verify its validity from the facts of biological and historical evolution. Thus, although the core message of this book is spiritual, I would like to encourage the reader to look upon its facts and logic as he or she would with any scientific theory. The common criteria for a valid scientific theory are that it should be empirically verifiable and explain the widest possible range of phenomena in the simplest possible way. As we apply the Mayan chronology to human history, a significant wave pattern becomes visible. Examples of such patterns presented in this book are the "winds of history," the emergence of religions, the development of written communication, and the ups and downs of the world economy. These examples clearly demonstrate that we are living in a conscious universe whose built-in intelligence follows an exact schedule according to which humans are meant to evolve. In this schedule we have now reached the semifinal Underworld (the eighth of nine basic Mayan time cycles), which means that the destiny of humanity needs to be fulfilled in less than ten years. We are truly living in the eleventh hour of creation!

The degree to which the facts of modern science and historical research match up with the Mayan calendar is by most standards astounding. I am convinced many readers will conclude that this is the most important source of knowledge about the cosmic plan available to humankind, which leads to the further conclusion that the dissemination of the Mayan calendar in its true—that is, empirically verifiable—form today is an urgent task for all of us. Spiritual books often repeat the idea that there are many different paths, and there is no right path for everyone. While this is true, it is too seldom recognized that it is true only in a narrow individualist sense. The only reason that there are any "paths" to begin with is that there is a cosmic time plan that governs the evolution of consciousness and propels us all to evolve on our paths toward a common destiny. If our individual paths are not seen in the larger context of this plan, they completely lose their meaning. Today we as a species are faced with the challenge of fitting our individual paths within this larger plan so as to be able to meet its deadline. The Mayan calendar, then, is of utmost concern to all of us.

This situation presents spiritual seekers with a choice. We may continue to use astronomically based calendars, and, if we do, we will continue to see history as a series of chaotic random events. Alternatively, we may begin to use the traditional Mayan calendar, become aware of the contours of the cosmic plan, and align ourselves with its evolution of consciousness. An important consequence of this ongoing evolution is that the idea of a fixed "human nature" lacks any basis in reality. Differences in consciousness between people living in different Underworlds are real, and human consciousness today is not the same as it was for the ancient Maya. Nor is it the same as that of the Universal Human Being of the forthcoming golden age. There is hope for humanity, not because we will all suddenly choose to change for the better, but because the consciousness of humanity is subject to a cosmic plan that cannot be manipulated. To be immersed in this plan through the use of the Mayan calendar will further its fulfillment on both an individual and a collective level.

Moreover, the shifting yin/yang polarities of this plan explain why the world is looked upon differently not only in different eras, but also in different corners of the world.

The current time well exemplifies this latter point. As these words are written in the seventh *uinal* (twenty-day cycle) of the fifth *tun* of the Galactic Underworld, the West once again seems to have been successful in one of its military campaigns against an Islamic nation of the East, Iraq, where the Saddam Hussein regime has been toppled (more on this in appendix D). It is natural to ask the reason for this war. The mainstream media has been touting superficial answers: the United States wanted to get rid of an oppressive regime that may have possessed weapons of mass destruction; or, alternatively, the United States wanted to take control of Iraq's oil resources. But such answers are given within the framework of the ruling consciousness and are consequently very shortsighted.

Suppose instead that the currently emerging shift in the relationship between the Old World and the New (in the geographical sense) is not at all a function of differing viewpoints on the war in Iraq. Suppose that the larger shift was something that was meant to happen because of a change in consciousness that is scheduled to take place on a global level. Thus, this war, occurring in what I have called the Third Day of the Galactic Underworld, may not be remembered primarily as another demonstration of the military prowess of the sole Western superpower. What may be of more profound consequence is that the system of international law has been toppled. The neglect by the Western coalition of the Security Council had as its outcome the de facto collapse of the United Nations and even of NATO. It is hard to believe that the West will ever again be able to muster support for the world order it heads other than through force.

We saw an opposite tendency of equal importance in the period prior to this war, a global movement for peace of an unprecedented scope. This is a reflection of the fact that many people in the approach to the golden age are increasingly living in the present and unwilling to postpone its fruits.

Was all this predictable? The idea of an intensified conflict between the Old World and the New World may certainly ring a bell for those familiar with Hopi prophecy. Moreover, the student of my previous book, *The Mayan Calendar,* would hardly be surprised by the emerging split between the United States/Great Britain and continental Europe. As was explained in *The Mayan Calendar,* the period from 1992 to 1999 was one of the most peaceful in the history of humankind, while I predicted that the period beginning in early 1999 would become an era of increasing conflicts between East and West. In fact, I described this whole Underworld as the era that in Christian terminology is called the Apocalypse, whose Beast may then—paradoxically, perhaps—primarily come to be constituted by those who tend to project evil onto others.

There is much more to learn from the prophetic science of time regarding the further development of the current Underworld, and these matters are discussed extensively in the later chapters of this book. This eighth and Galactic Underworld has, for instance, in contrast to the period from 1992 to 1999, been one of a steady economic decline, and there are reasons to expect that the world will never again experience a prolonged period of economic growth (see appendix A).

The key to understanding that prophecy is indeed possible lies in the recognition that human thinking is not something that takes place "inside" the head of a person in isolation from the rest of the cosmos. Our thinking, and by consequence our actions as well, develop largely through resonance with an evolving cosmic consciousness mediated by the earth, whose various energy shifts are described by the Mayan calendar. Thus, we are all "channelers"; and, frankly, at least until now, we have been able to think for ourselves in only a very limited sense. Through our resonance with the earth we have all been more or less puppets of the cosmic plan. There is, then, nothing mysterious about the fact that the large-scale development of this world of puppets is essentially predictable. Prophecy is, in other words, fully possible, and to understand the patterns of the cosmic plan is a very important task for all of us.

A great value of the Mayan calendar is that it provides us with knowledge of the energies that guide evolution. It is a tool that enables us to go with the flow. Although we have been unconscious of it, such guidance has always been available to human beings and will be so until the completion of the cosmic plan on the day 13 Ahau, October 28, 2011. Yet for those who want to move away from being puppets and seek instead to become conscious co-creators, this book makes possible an awareness of the energies driving evolution—an awareness that has not been fully available since the days of the ancient Maya.

The traditional Mayan calendar is now being revived, and its followers will be its modern exponents. Although initially the reader is prepared for this with some basic knowledge regarding the structure and meaning of the Mayan calendar, I have written the latter part of this book as a guide to where the divine plan is meant to take us: the golden age.

Ruling Deity Time Span	Historical Dates	A Brief Timeline of Mesoamerican Cultures
Xiuhtecuhtli 3115–2721 BCE	3000 BCE	First cultivation of maize
Tlaltecuhtli 2721–2326 BCE		
Chalchiuhtlicue 2326–1932 BCE		
Tonatiuh 1932–1538 BCE		
Tlacolteotl 1538–1144 BCE	1500 BCE	Olmec civilization in Veracruz
Mictlantecuhtli 1144–749 BCE		
Cinteotl 749–355 BCE	550 BCE	Zapotecs in Oaxaca First Tzolkin date in Monte Alban
Tlaloc 355 BCE–40 CE	250 BCE–50 CE 32 BCE	Preclassical Maya, Izapan culture First Long Count date
Quetzalcoatl 40–434 CE	100 CE	Beginning of classical Mayan culture Rise of the city of Teotihuacán in central Mexico
Tezcatlipoca 434–829 CE	434 CE 700 CE 800–830 CE	Founding of dynasties in Copán and Palenque Collapse of Teotihuacán Collapse of classical Mayan culture
Yohualticitl 829–1223 CE	843 CE 909 CE 1223 CE 1223 CE	Earliest date in Chichén Itzá Beginning of postclassical Mayan culture Last use of Long Count date Collapse of postclassical Maya in Chichén Itzá Collapse of Toltecs in Tula
Tlahuizcalpantecuhtli 1223–1617 CE	1368 CE 1504 CE 1519 CE 1521 CE	Aztecs settle in Mexico Valley Columbus spies Mayan canoe off coast of Honduras Cortés lands in Veracruz; Fall of Aztec empire
Ometeotl 1617–2011 CE	1697 CE	Fall of last independent Mayan kingdom at Tayasal

Figure 1.1. A brief timeline of Native Americans in Mesoamerica

1.

The Legacy of
the Maya

NATIVE AMERICA

When we first acquaint ourselves with the myths, legends, and
cosmological ideas of the ancient Western Hemisphere they
may seem bewildering. The Mayans and the Aztecs described a
universe of Thirteen Heavens and Nine Underworlds with
which we, with our modern mind-set, find it difficult to identify. The first reaction of many is to see such ideas as supersti-
tious and belonging to the prescientific worldview of people
who did not know any better. Yet the Maya, especially, do not
cease to fascinate us. Deep down, many have an intuitive feeling that this people possessed a deeper wisdom than that of
our own modern civilization. Time after time, the ancient
Maya come to our attention, and they will probably continue
to do so until we learn what there is to learn from them. The
overwhelming majority of people, however, lack the proper
preparation to understand this ancient Native American
worldview and so are looking in directions where no understanding is to be found.

How then are we to prepare ourselves to assimilate the

1

ancient Native American knowledge? The answer is simple: through a study of the Sacred Calendar, or *tzolkin*. The Sacred Calendar is the main entry to the thinking of the advanced civilizations that existed in the Western world before the arrival of the Europeans. By retrieving the Mayan calendar after centuries of oppression of its indigenous populations, we may recreate the piece of the cosmic puzzle provided by Native America. The Sacred Calendar provides the key to the unification of the widely varying perspectives in today's world on what it means to be a human being and gives us a tool for aligning our intuition with our individual and cosmic purposes.

In ancient times the whole region that stretches from northern Mexico to Honduras formed a single cultural context that archaeologists usually call Mesoamerica (fig. 1.2). Although the Maya, Zapotecs, Mixtecs, Toltecs, Teotihuacanos, and later Mexicas (commonly known as Aztecs) were mostly politically separate, artistic and other influences spread all over this region (see the inside cover for a brief timeline). Contacts through merchants and others were intense, and these cultures were all interrelated. Within this Mesoamerican civilization people seemed well aware of neighboring regions, such as the Caribbean and North America proper—then known as Turtle Island—and South America.

What is the legacy of this ancient civilization? What have modern people inherited from its knowledge? Nothing, it seems, and few would assert that any part of the Native American knowledge is crucial to our current understanding of the world. Although we know of the reverence felt by Native Americans toward nature, we lack the tools to recreate their oneness with all things living. It is almost as if half the world, the Western world, has had no influence on today's global culture. Even in Mexico itself, where almost half of the world's population of Native Americans now lives, the ancient spiritual knowledge is little known.

Why is this so? The most immediate explanation is the near annihilation of the indigenous populations of the Americas and of their cultures. When the Spanish conquistador Hernán Cortés landed on the east coast of Mexico on Good Friday, 1519, for example, the total pop-

Figure 1.2. Map of the Mesoamerican region showing both the areas of ancient cultures and modern national borders

ulation of Mesoamerica was an estimated 25 million people; a century later it had fallen to about 1 million. Even if this decline in numbers was only partially the result of direct massacres and was primarily caused by the diseases carried by the invaders, the outcome was the same: a civilization was annihilated.

This onslaught also meant cultural destruction. The Spanish bishops burned all the books they could find that had been written by the Maya, a people that was alone in the Western Hemisphere in possessing a written language and carried many of its most advanced cultural expressions. Today only four of these books remain. They escaped the destruction only because they were sent to Europe at a very early point; much later they surfaced in libraries there. The surviving Mayan books are all calendars (fig. 1.3).

But the burning of books was only the most visible way in which the old civilization was destroyed. The conquistadors and the friars forbade all forms of the traditional religion and way of life. In many areas, such

Figure 1.3. A page from the Venus tables of the Dresden Codex, which is considered to be the finest Mayan calendrical work to have survived the book burnings by the Spanish friars

as Guatemala, the persecution of the Maya has continued up until the present time, and much of the traditional knowledge has been driven underground. Under these circumstances, it is not surprising that very little of the knowledge possessed by the natives has become known and that most people do not even know that it exists. Today most people take it for granted that if the Maya had some knowledge that was really important to the modern world, we would know about it. But is this necessarily true? Could the archaeologists and anthropologists of today really tell what is of value and what is not? I would say no. Today's university scientists consider modern science as the norm and judge the advancement of Native American civilization by modern standards. If we are to recognize the true value of Native American science, we must do it the other way around: evaluate the quality of modern knowledge using the ancient knowledge as a standard. Only from such a point of departure will we realize that humanity has lost access to an indispensable part of its ancient heritage, a heritage that has either been distorted or not understood.

This very lack of insight about the knowledge of Native America has created a global imbalance. Half the world has been shut out. Today's global culture is often called Western, but in the global culture the real Western contribution, its ancient original thinking, is missing. This global imbalance has led to a general lack of understanding of the purpose of life and what it means to be a human being. Labeling the global culture "Western" has blinded us to the fact that the *native* Western viewpoint is the badly needed piece of the puzzle. We need this piece to be able to seal the frame of a holistic understanding of the world, and in the absence of this piece we allow ourselves to be deluded by a false worldview.

I do not, therefore, present the knowledge of the ancient indigenous civilizations as something of the past. This book does not look upon the knowledge of the ancient Mesoamericans as a curiosity with which to indulge our fantasies or as a reason to tease ourselves with unsolved mysteries. Rather its purpose is to show that they were right and we are wrong, at least when it comes to the big picture. This book asserts that the ancient knowledge is knowledge for the future, knowledge that cements the truth to which the world now must awaken.

THE HISTORICAL MAYANS

It is important to be aware of the existence of a wider common framework of Native America prior to the advent of the Europeans, since this civilization is often described as a multitude of more or less isolated tribes. It is also important to understand the central place of Mesoamerica in this framework. Even today Mexico has a native population that is estimated to be more than ten times greater than that of the United States. So if this book focuses mostly on the science and calendar of the Maya, it is because I regard these as the most clearly articulated and exact expressions of a worldview that in important aspects was shared by a whole continent.

Before describing the calendrical system of the Maya, I will give a very brief history of the people who developed it. The high culture of the Maya flourished during the first millennium after Christ and was embraced by about 5 million people living in city-states in the area that is currently southern Mexico, Guatemala, and Belize (fig. 1.4). Within this area the northern Yucatán peninsula is geographically a very flat lowland area, whereas Chiapas and Guatemala are jungle areas partly located in the mountains. The most important foods are maize, beans, and jalapeños; rubber, vanilla, coffee, and cocoa are cultivated as well.

Except through their art and architecture in pyramid sites such as Palenque, Tikal, Copán, and Chichén Itzá, the classical Mayans are best known for their astronomical observations and mathematical knowledge. The Maya were the first people on earth to make use of the number 0 and were clearly the most advanced astronomers and mathematicians of their day.

Over the years, there has been much speculation about the origin of the Maya, who have been variously portrayed as one of the lost tribes of Israel or as originating from China or Atlantis. This line of thinking is a continuation of that of the Spanish colonizers, who could not imagine that it was the Mayans themselves, the people who still inhabited the area, who had once built the great temple cities in the jungles. In recent times, however, research into the ancient Maya has entered a new phase. It is now possible to read most of their glyphic symbols on stone lintels and stelae, because it has become obvious that their language was an early form of the languages still spoken in the area.

As mentioned, only four written codices of the Maya from pre-Columbian times are known to have survived the moisture of the jun-

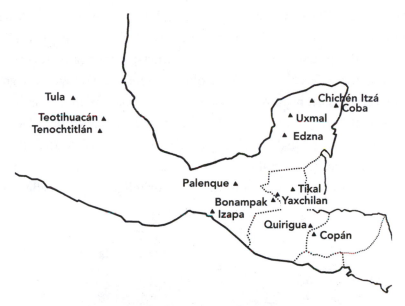

Figure 1.4. Map of Mesoamerica and the Mayan region with some of its most important ancient sites

gle and the zeal of the missionaries. But some books written down after the arrival of the Spanish, such as the Books of Chilam Balam (Books of the Jaguar Prophet) and the Popol Vuh, have also provided information about the ancient mythology and cosmology of the Maya. The Books of Chilam Balam are prophetic books; the Popol Vuh has been described as the "Bible" of the Maya, including a creation tale in which the Hero Twins Hunahpu and Xbalanque outsmart the lords of death.

From these sources we know that the Maya believed in an afterlife and in a rebirth of the soul. Life was seen as a cycle of birth, death, and rebirth, and sacrifice was an important part of the creation of new life. The Maya had a dualist religion that centered on a struggle between good and evil, including good and evil gods. Examples of gods are Chac, the rain god; Yum Kax, the maize god; and Kukulcan, the feathered serpent, gods who were seen as animating the many aspects of human existence. In all their undertakings the Maya interacted with the gods.

During the so-called classical period, usually set as 250–900 C.E., the different city-states were governed by dynastic lineages of *ahauob* (kings who were also leaders of shamanistic rituals), who, in the eyes of their peoples, embodied the cosmos. The tasks of these shaman-kings included the public performance of sacrificial offerings of their own blood on the different pyramids, evoking a state of union with the gods and receiving visions for guidance. Much information about the life of the ancient Maya, or at least of their kings, is available from stelae and lintels describing rituals or glorifying winners in dynastic struggles or wars. Many such stelae also mark transitions from one time cycle to another, which were important occasions for religious rituals (fig. 1.5).

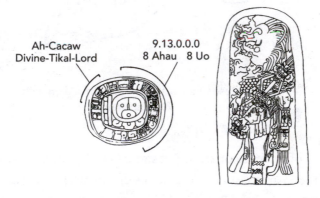

Figure 1.5. The king Ah-Cacaw celebrating the beginning (9.13.0.0.0 8 Ahau) of a new *katun* in Tikal (692 C.E.). The Mayan shaman kings are often depicted on stelae celebrating the beginnings of new sacred cycles of time on Ahau days.

In a technological sense, however, the Maya never left the Stone Age. Except in toys, they do not appear to have made use of the wheel, for instance. Yet in architectural style and art the Mayan centers display a great and impressive variation: Tikal is massive, and at its height is believed to have included some twenty thousand buildings, among which are many pyramids that once were painted red. The Yucatec centers of Chichén Itzá, Uxmal, Edzna, and Mayapan were laid out more openly. Palenque, with its palace and many small, almost Chinese-style, temples is exquisite and magical, situated on a jungle slope facing the plains.

The Mayan pyramids were sometimes laid out in alignment with astronomical events such as the rising of the Pleiades or the spring equinox—as in the famous case of the Pyramid of Kukulcan in Chichén Itzá. Of great symbolic importance, the many central pyramids, such as the Pyramid of Kukulcan, the Temple of the Inscriptions in Palenque, the Great Pyramid in Uxmal, and the Pyramid of the Jaguar in Tikal—the highest pyramid in the Americas (fig. 1.6), rising 144 feet (44 meters) above ground level—are all built in nine levels. These terracelike buildings reflect how the Nine Underworlds construct the cosmos.

About 800–830 C.E., the Maya abandoned almost all their city-states in Guatemala and Chiapas. This event did not take place simultaneously everywhere; over a period of about thirty years, one after another of the city-states that had been inhabited for centuries ceased to erect commemorative inscriptions and were evidently deserted. At around the same time, however, a new Mayan culture, centered on the location of Chichén Itzá, emerged at the northern Yucatán peninsula. This is a culture that is mostly referred to as postclassical.

This postclassical culture seems to have downplayed the role of shaman-kings and instead emphasized the cult of Kukulcan, the Feathered Serpent, better known by its Aztec name of Quetzalcoatl. This culture also came to an end, however (for reasons that are not easily understood from the perspective of traditional historical research), and by around 1220 C.E., Chichén Itzá was apparently deserted. The later descendants of the classical and postclassical Maya then had their first contact with the Europeans in the form of Christopher Columbus's fourth journey in 1504. Following the Spanish conquest of the Aztec empire in 1521, they were subjected to a series of invasions that began with Hernán Cortés's Honduran expedition in 1525 and continued until their last independent kingdom fell in 1697.

In many ways the present-day Maya have carried with them the traditions from the classical period, although over time many of these traditions have been modified under the influence of the Europeans and Christianity. The Lacandon group in the rain forest of Chiapas still lives traditionally and sacrifices to some of the ancient gods at the old temple sites. Over the past several decades the Maya

Figure 1.6. The nine-story Pyramid of the Jaguar in Tikal, the tallest pyramid in the Americas at 144 feet. Its nine stories are symbolic of the nine levels of consciousness. Climbing to the highest level of the pyramid to perform rituals, the Mayan shaman kings symbolically reached the cosmic level of consciousness.

population in Guatemala has suffered from intense repression by the various military regimes that followed the coup against Jacobo Arbenz in 1954, and tens of thousands have been murdered.

THE RETURN OF THE SACRED CALENDAR

Almost everyone who has studied the ancient civilizations of Mesoamerica agrees that the road to understanding their cosmology—the way these peoples looked upon the world—runs through their calendar. Clearly, the Maya were the people who had developed the most advanced calendrical system. Until the crushing of the Mexicas, the entire region, reaching into the current United States, shared the core of this calendar system, the 260-day Sacred Calendar. Before the arrival of the Spanish, the whole region was spiritually synchronized through the use of this same calendar.

It may seem odd that a calendar could be this important. Most would think the choice of a calendar is rather arbitrary and has little influence on the worldview of a civilization. But perhaps it is exactly because the choice of calendar appears so unimportant, like a mere convention, that it has become a blind spot in our consciousness. The Gregorian calendar, instituted by Pope Gregory in 1582 and now in use worldwide, is taken as a given, and there seems little reason to replace or complement it.

But what if this calendar exerts an insidious indoctrination that most people are unaware of and strengthens a worldview that is false? This book is written partly to highlight this indoctrination and hold up the true Sacred Calendar as an alternative. It is as an alternative to the worldview imposed on us by the Gregorian calendar that the traditional Mayan calendar is now reemerging from the ashes. This Sacred Calendar is a codification of the Mayan Universe of Holy Time.

To make things worse, the Mayan calendar has been largely misrepresented by regular tourist guides and textbooks. Typically, standard books emphasize how advanced the Mayan sciences of mathematics and astronomy were in ancient days. While this may be true, it is almost entirely beside the point and by itself would not be a good cause for the revival of the Mayan calendar. After all, the astronomical measurements of today's scientists are much more accurate than those of the ancient Maya, and if astronomy was all there was to it there would be no good reason for us to invest a deeper interest. What is important to today's world are not the astronomical aspects of the Mayan calendar but the spiritual. Through the spiritually based nonastronomical calendars, the *tun* (360-day period) and the tzolkin (260-day period), the true, and unparalleled, contribution of the Maya to modern humanity is to be found.

The tzolkin, meaning "count of days" in Yucatec Maya, is also known as the Sacred Calendar. In Guatemala, the Quiché-Maya people have kept this calendar intact for a period of 2,500 years through the diligent observance of the calendar by so-called day-keepers, men and women endowed with the special responsibility of keeping track of the days. Through comparisons with dates on old stelae, archaeologists have been able to verify that not a single day has been lost in 2,500 years.

The Mayan calendar is still being used. In the past two decades, interest in the Mayan calendar has increased worldwide, and the work to reconstruct it and develop a true spiritual calendar charting the future of humanity has begun. This revival is occurring partly among researchers, who take as a point of departure that the worldview of the Maya was, and is, closer to the truth than the modern version. It is also happening among the living Maya themselves, many of whom are educating their young in the old ways. At the core of the teaching is the Sacred Calendar.

THE COUNT OF DAYS: THE TZOLKIN

In the Sacred Calendar the days are counted differently from calendars developed in other parts of the world. In this calendar the days are counted in two ways. The numbers 1 through 13 (fig. 1.7) constitute a thirteen-day count or "thirtnight" (English has no word for it, but Spanish does: *trecena*). In parallel with this, each day is assigned one of twenty different signs that are always counted in a specific order, resulting in a twenty-day cycle the Maya call a *uinal*. The two counts, the trecena and the uinal, run together so that each day is characterized by both a number and a sign. A common way of illustrating this is by two cogwheels, one with the thirteen numbers and the other with the

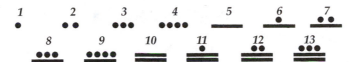

Figure 1.7. The numbers 1–13 in Mayan writing.
A dot represents 1 and a bar represents 5.

twenty day signs, each of which moves one step every day to produce a new combination (fig. 1.8).

As every day is characterized by both a number and a sign, 13 × 20 = 260 different combinations of numbers and signs are generated in order. The first day is 1 Imix, the second 2 Ik, then 3 Akbal, 4 Kan, 5 Chicchan, 6 Cimi, 7 Manik, 8 Lamat, 9 Muluc, 10 Oc, 11 Chuen, 12 Eb, 13 Ben, and then not 14 but 1 Imix, because the thirteen numbers start on a new round. Each such combination of a number and a glyph recurs every 260 days. The chart of these 260 combinations of the thirteen-day count and the twenty day signs, which is shown in figures 1.9–1.10, was

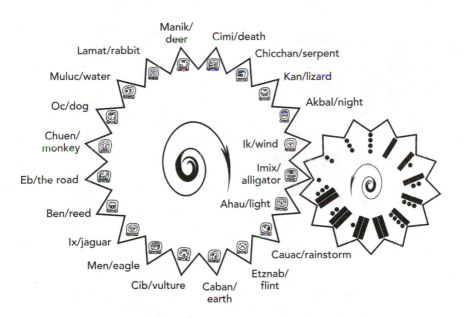

Figure 1.8. Rotating cog model of how the different tzolkin combinations (here, 1 Imix) are generated from the twenty day signs and the thirteen numbers

Tzolkin Chart 1

Mayan Day Signs   **Aztec Day Signs**

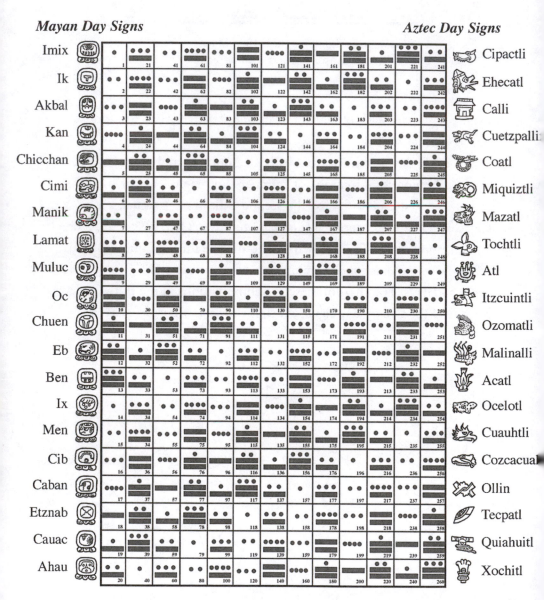

Figure 1.9. Tzolkin chart showing the twenty Mayan day signs at the far left and the corresponding Aztec day signs at the far right. Also shown are the thirteen numbers with which the day signs combine. Each day should be read in the order of its day number, or kin, in the lower right corner of each number, or as in Tzolkin Chart 2 (fig. 1.10).

Tzolkin Chart 2

Mayan Day Signs   **Aztec Day Signs**

Mayan Day Sign														Aztec Day Sign
Alligator	1	21	41	61	81	101	121	141	161	181	201	221	241	Alligator
Wind	2	22	42	62	82	102	122	142	162	182	202	222	242	Wind
Night	3	23	43	63	83	103	123	143	163	183	203	223	243	House
Seed	4	24	44	64	84	104	124	144	164	184	204	224	244	Lizard
Serpent	5	25	45	65	85	105	125	145	165	185	205	225	245	Serpent
Death	6	26	46	66	86	106	126	146	166	186	206	226	246	Death
Deer	7	27	47	67	87	107	127	147	167	187	207	227	247	Deer
Rabbit	8	28	48	68	88	108	128	148	168	188	208	228	248	Rabbit
Water	9	29	49	69	89	109	129	149	169	189	209	229	249	Water
Dog	10	30	50	70	90	110	130	150	170	190	210	230	250	Dog
Monkey	11	31	51	71	91	111	131	151	171	191	211	231	251	Monkey
The Road	12	32	52	72	92	112	132	152	172	192	212	232	252	Grass
Reed	13	33	53	73	93	113	133	153	173	193	213	233	253	Reed
Jaguar	14	34	54	74	94	114	134	154	174	194	214	234	254	Ocelot
Eagle	15	35	55	75	95	115	135	155	175	195	215	235	255	Eagle
Vulture/Owl	16	36	56	76	96	116	136	156	176	196	216	236	256	Vulture
Earth	17	37	57	77	97	117	137	157	177	197	217	237	257	Movement
Flint	18	38	58	78	98	118	138	158	178	198	218	238	258	Knife
Rainstorm	19	39	59	79	99	119	139	159	179	199	219	239	259	Rain
Light/Lord	20	40	60	80	100	120	140	160	180	200	220	240	260	Flower

Figure 1.10. Tzolkin chart showing the twenty Mayan and Aztec day signs with their English meanings. Also shown are the 260 kin numbers, which can be cross-referenced to Tzolkin Chart 1 (fig. 1.9).

referred to as the *tonalpouhalli* by the Mexicas and the tzolkin among the Yucatec Maya. It is still regarded as the Sacred Calendar among the living Maya, reflecting a process of divine creation that proceeds without interruption.

Although the Maya and the Mexica used different day signs (fig. 1.9) and had different names for them, they had essentially the same meanings. A day that was 4 Manik in the Yucatán would be 4 Mazatl among the Mexica, but in both cultures this would mean 4 Deer. The days were believed to be ruled by different deities, or "energies," and thus they were symbolized by different glyphs. Every day had its own energy, or Day Lord, and throughout the region there was, with some variations, agreement about these. The meaning of these Day Lords is further discussed in appendix B.

Why did the ancient Mexicans count time in this particular way? Why, despite the pressure to conform to the Gregorian calendar, have the Maya kept it in existence to this day? And why is this calendar considered sacred? This book seeks to answer these questions. Time and timekeeping are blind spots in our modern culture, so the answers proposed will seem surprising to many. Initially most people would probably not think it matters whether they count the days by a seven-day week combined with a month of twenty-eight, thirty, or thirty-one days, as in the Gregorian calendar, or by thirteen days and twenty glyphs, as in the Sacred Calendar. But I hope the reader will come to agree that it is a very important difference indeed, and that the way we count the days has a profound influence on our worldview. In fact, it deeply affects our ideas about what it means to be a human being.

Where to start? With the number 13!

2
··

The Thirteen Heavens

THE LONG COUNT AND ITS BEGINNING

To understand the Sacred Calendar and its deeper meaning, it is easiest to begin by taking a broader view of the calendrical system of the Maya.

The Long Count is the name given to the chronology used by the Maya in the classical era to keep track of the long-term passage of time. On almost all the ancient pyramids and stelae, dates were inscribed according to this Long Count. The Long Count consisted of thirteen *baktuns,* which are periods of 400 tuns (360-day periods). One baktun is thus 400 × 360 = 144,000 days, amounting to 394.3 solar years.

Today most archaeologists agree that the beginning date of the thirteen baktuns of the Long Count was the day we would now call August 11, 3114 B.C.E. If we add thirteen baktuns to this date we arrive at December 21, 2012 C.E., the date that has become known as the end of the Mayan calendar. This date has aroused the interest of people all over the world because it is very close to our present time, and much discussion and specu-lation have focused on what will happen then.

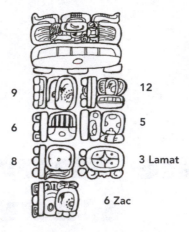

Figure 2.1. The full classical-era description in Mayan hieroglyphics of a Long Count date (9.12.6.5.8 3 Lamat). At the top is the glyph signifying that this is a Long Count date, and below it are the various numbers of baktuns, katuns, and so forth. The glyph at the bottom specifies the day in the *haab* (agricultural) month, 6 Zac.

Before we can use the Mayan calendar system prophetically, we must first demonstrate its ability to describe the past. The Long Count of thirteen baktuns is parallel to the trecena, or count of thirteen days within the tzolkin. An important common factor in the two calendars is thus the count of thirteen, a number that played a crucial role in many ancient civilizations. It was considered holy not only in ancient Mexico but in religious systems all over the world. In the Christian religion, the number 13 played a noteworthy role in that Jesus and his disciples formed a group of thirteen members.

Since the Maya began to use the Long Count (around the time that Quetzalcoatl, the creator of the calendar, took up his rule in the Heaven) (fig. 2.5, page 23), at about the same time that Jesus gathered his twelve disciples, perhaps people in different parts of the world simultaneously came to realize there was something special about the number 13. All over ancient Mesoamerica people shared the view that there were thirteen Heavens. In more recent times, however, thirteen has been regarded as a number signifying bad luck. Regardless of how we interpret its meaning, there seems always to be a charge on the number thirteen.

Day No.	Ruling Aztec Deity	Associated Bird	Growth Stage
1	**Xiuhtecuhtli,** god of fire and time	Blue hummingbird	Sowing
2	**Tlaltecuhtli,** god of earth	Green hummingbird	
3	**Chalchiuhtlicue,** goddess of water	Falcon	Germination
4	**Tonatiuh,** god of the sun and warriors	Quail	
5	**Tlacolteotl,** goddess of love and childbirth	Hawk	Sprouting
6	**Mictlantecuhtli,** god of death	Owl	
7	**Cinteotl,** god of maize and sustenance	Butterfly	Proliferation
8	**Tlaloc,** god of rain and war	Eagle	
9	**Quetzalcoatl,** god of light	Turkey	Budding
10	**Tezcatlipoca,** god of darkness	Horned owl	
11	**Yohualticitl,** goddess of birth	Scarlet macaw	Flowering
12	**Tlahuizcalpantecuhtli,** god before dawn	Quetzal bird	
13	**Ometeotl/Omecinatl,** Dual-Creator God	Parrot	Fruition

Figure 2.2. Deities and evolutionary symbols associated with each of the thirteen days/Heavens of the trecena, defining their spiritual qualities. Each day in the thirteen-day count was considered to be a step in a process of growth and was symbolized by a bird. As the thirteen days progress, these birds symbolize an evolution from those that are small—such as the hummingbird—to those that are more spectacular, such as the quetzal or the parrot.

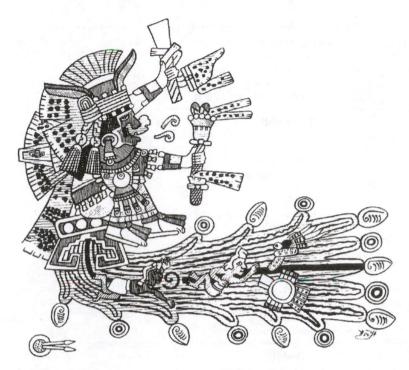

Figure 2.3. Chalchiuhtlicue, the Aztec goddess of water and ruler of the third day of the trecena. She is seen here giving birth to infants, which may be seen as an act symbolic of the quality of the day she rules.

Where did this idea of thirteen Heavens come from? What was its basis? A first clue is that the thirteen-day count of the tzolkin in ancient times was seen as a reflection of a process of creation, the unfolding of evolution from seed to mature fruit, which took place in thirteen steps (fig. 2.2). Each of these steps was also brought about by a certain deity. Since the names used by the Maya for the deities ruling these Heavens are no longer known to us, in what follows we use the names given to them by the Aztecs.

In the ancient Mesoamerican Universe of Holy Time, every day was ruled by a special god and symbolized by a bird with magical qualities. The energy, or divine power, of each of these gods corresponded to that of the one of the Thirteen Heavens which he or she ruled. Mesoamerican gods are very multifaceted and take part in a wide range of stories, however, so it is not always easy to know what they were meant to symbolize in their guises as Lords of Time. From a study of Aztec mythology in other sources, the interested reader may learn more about the nature of the gods and energies ruling the Thirteen Heavens. What is crucial to understand here is that these Thirteen Heavens dominated creation processes on many different levels, not only the thirteen days of the tre-cena, but also, for instance, the thirteen baktuns of the Long Count.

An overall impression given by figure 2.2 is that the gods ruling the odd-numbered Heavens seem more nurturing and female than the gods ruling the even-numbered Heavens, who are mostly warlike and male. It should be pointed out that the deities ruling the First, Seventh, and Thirteenth Heavens were regarded as one and the same deity in differ-ent guises. Xiuhtecuhtli is the Creator God's manifestation on earth—the navel of the world—whereas Ometeotl is his aspect in the highest realm above. Cinteotl, the god ruling the Seventh Heaven, is the same deity, but in his aspect of the maize god. These three deities, who are one and the same, are all dual gods unified with female companions.

Did these "deities" really play a role in human life? The important question we now face is whether they actually imposed their energies on the Thirteen Heavens, or whether, as most people today would probably surmise, they were just superstitions. To find out, let us as an experiment forget the worldview that has been imposed on us by our schools, universities, churches, and television programs and explore reality with the Mesoamerican calendar as a guide.

THE EMERGENCE OF HIGHER CIVILIZATIONS

We start by seeing how these Thirteen Heavens have manifested in the creation process of the thirteen baktuns of the Long Count. In such a large-scale historical perspective, we may be able to see features that are not so easily evident in a sequence of merely thirteen single days. Yet the beginning date of the Long Count, August 11, 3114 B.C.E., has tra-ditionally been considered by archaeologists as a "mythic" date with no real meaning. In recent decades, however, some researchers have begun to see this date differently. If we consider the Mayan calendar as a calendar of the whole planet rather than of an isolated culture, we discover new and interesting things.

It is well known that very important events, some of the most important in the history of humankind, took place on earth at the time of the beginning of the Mayan Long Count. The upper and lower parts of Egypt, for instance, were unified into one nation about 3100 B.C.E. King Menes, who accomplished this, became the founder of the First Dynasty of Egypt and, as a sign of his new power, the first wearer of the Pharaonic Double Crown. Thus Egypt, the world's first nation, emerged and came to be ruled by its first pharaoh very close to the time when the Mayan Long Count begins. At around the same time the first larger monumental buildings on this planet were also built. The oldest pyramid in Egypt, Pharaoh Djoser's pyramid (fig. 2.4), has been dated by carbon 14 techniques to about 2975 B.C.E. Stonehenge in England has been dated somewhat earlier, as have the Sumerian pyramids and the large Newgrange structure in Ireland. Also, the effective use of metals in the form of bronze began in Crete and Sumer, so at this very time humankind took the first step out of the Stone Age.

Figure 2.4. One of the world's oldest large-scale constructions, Pharaoh Djoser's pyramid (c. 2975 B.C.E.) in Saqqâra, Egypt, was originally built in seven stories.

But there is more to it. The Sumerians, who lived in present-day Iraq and are often considered to have built the first higher civilization on this planet, had begun the use of writing only slightly earlier than the beginning of the Long Count, about 3200 B.C.E. The beginning of what we define as the historical era—the period when humanity has had written language—thus coincides with that of the Mayan Long Count. When the First Heaven of the Long Count began its rule on August 11, 3114 B.C.E., human history also began.

The ancient Egyptians regarded the unification of lower and upper Egypt, the birth of their own nation, as an act of divine intervention. The Egyptians saw the beginning of the Mayan Long Count as a time of cosmic creation brought about by the gods, as the Maya also did later. Is this concordance of the Egyptian and Mayan cosmologies truly an accident, or could there be some kind of cosmic creation plan in operation that governs historical processes on earth?

To find out, let us first look at the deity who presided over this surge of creativity at the beginning of the Long Count. According to the Temple of the Inscriptions in Palenque, the First Heaven of the Long Count was established by the First Father, who then raised the World Tree and so established the Four Directions. According to the Mexicas (fig. 2.5), the god of fire and time, Xiuhtecuhtli, who had both a male and female aspect, ruled this First Heaven. The views of the two Mesoamerican peoples are not contradictory, however. Like the First Father among the Maya, Xiuhtecuhtli was a god of procreation who had existed since the beginning of time and brought light into the darkness.

The suddenness with which the first higher human civilizations arose some five thousand years ago has long baffled historians. But if we are willing to accept that a cosmic time plan exists, this sudden emergence is not surprising at all. The emergence of human civilization is, in fact, best explained by the energies of time exemplified by the deities of the Mayan calendar. The sudden emergence of the first higher human civilizations may then be regarded as the first step in an evolutionary process that has continued throughout the thirteen baktuns of the Long Count. The first of these Thirteen Heavens only planted the seeds for a human civilization

Ruling Aztec Diety	Heaven Number	Heaven Starts	Heaven Ends
Xiuhtecuhtli, god of fire and time	1	3115 BCE	2721 BCE
Tlaltecuhtli, god of earth	2	2721 BCE	2326 BCE
Chalchiuhtlicue, goddess of water	3	2326 BCE	1932 BCE
Tonatiuh, god of the sun and warriors	4	1932 BCE	1538 BCE
Tlacolteotl, goddess of love and childbirth	5	1538 BCE	1144 BCE
Mictlantecuhtli, god of death	6	1144 BCE	749 BCE
Cinteotl, god of maize and sustenance	7	749 BCE	355 BCE
Tlaloc, god of rain and war	8	355 BCE	40 CE
Quetzalcoatl, god of light	9	40 CE	434 CE
Tezcatlipoca, god of darkness	10	434 CE	829 CE
Yohualticitl, goddess of birth	11	829 CE	1223 CE
Tlahuizcalpantecuhtli, god before dawn	12	1223 CE	1617 CE
Ometeotl/Omecinatl, Dual-Creator God	13	1617 CE	2011 CE

Figure 2.5. The ruling deities of the Thirteen Heavens of the Great Cycle and their (corrected) durations in terms of Gregorian years

that, in a series of thirteen steps, has grown toward fruition. The Maya symbolized this process of growth by the steps of seven-story pyramids (fig. 2.6), such as the Pyramid of the High Priest in Chichén Itzá.

As we consider the emergence of the first higher civilizations on earth, we should note that this development also involved some negative

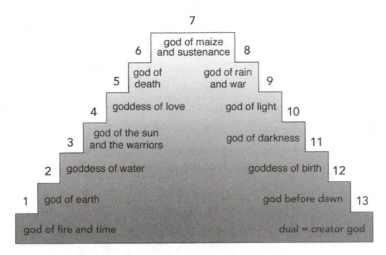

Figure 2.6. The progression through the Thirteen Heavens symbolized by the thirteen-step climb up and down a seven-story pyramid

aspects, which will become especially relevant when we later discuss humanity's current path toward enlightenment. The invention of writing and the construction of pyramids and monuments are usually viewed as advances on humanity's path, but these phenomena also indicate that after the onset of the Long Count human beings were no longer able to live in the present. Writing emerges only in a civilization that is undergoing change at such a high rate that there is a need to preserve information. Similarly, the Egyptian pyramids and monuments were built partly to immortalize the pharaohs; concern about leaving an imprint among mortals and making statements to future generations does not arise in a timeless culture. Perhaps the Stone Age people who had preceded the first higher civilizations were living more fully in the present.

Moreover, the emergence of deified rulers and a society divided by class indicates that at this point people began to think in terms of judgment, evaluation, and hierarchical ranking. A system in which one person ruled over many others replaced the more egalitarian, perhaps matriarchal, societies that had existed previously. The use of metal weapons and the first organized warfare are also proof that a mentality dominated by conflicts had emerged, one that still plagues us. The beginning of the Mayan Long Count meant, in fact, that a Fall had occurred, a

Fall that is spoken of in many different traditions. The beginning of the Jewish calendar only slightly earlier, in 3761 B.C.E., for example, marks the expulsion of the first human beings from the paradisiacal Garden of Eden. What if the Garden of Eden was a level of nondual consciousness, and its Tree of the Knowledge of Good and Evil was identical with the World Tree raised by the First Father of the Maya? But more on this later.

If this perspective—which establishes a link between the creative energy of the First Heaven of the Long Count and the emergence of human civilization—is true, the Maya, who invented a chronology to describe this process, must have had some profound insights about time and prophecy of which the rest of the world has been ignorant. In ancient Mesoamerica interest in time was always largely prophetic, and the Long Count represents the beginning of a cosmic creation cycle, which provided a foundation for prophecy. If this Long Count chronology holds information about human creativity in parts of the world other than Mesoamerica, it must also truly be a global calendar, of interest not only to the Maya but to us all. People around the globe are today intuitively attracted to the Mayan calendar for this very reason. This calendar has something to say about the direction in which life on earth is evolving; knowledge that is urgently needed at the present time.

THE HISTORY OF WRITTEN COMMUNICATION

To understand more of the nature of the spiritual energies of the Mayan calendar, we must explore how these have manifested throughout the course of human history. Only if we are convinced that the Mayan calendar describes a larger cosmic creation plan, shaping our consciousness and conditioning our creativity, does it become meaningful to learn what this calendar has to say about the future course of humanity.

So far, the only aspect for which we have found support is the idea that the first of the Heavens of the Long Count was ruled by a specific energy of creation, that of seeding. What were the results of the subsequent twelve Heavens of this cycle? To find out, we must first examine the time spans—the 394-year-long baktuns—during which the energies of these Thirteen Heavens of the Long Count dominated. (For reasons that are partly explained in chapter 7, these have been slightly altered from the classical Mayan form; the corrected cycle is hereafter called the Great Cycle). At this point the reader may pause to consider the time spans (fig. 2.5, page 24) that these Heavens cover and notice whether any important events in human history come to mind around the years that a new Heaven and deity have begun their rule. If the Mayan calendar is truly prophetic, we would expect great changes in the spiritual energy of the world at such shifts between baktuns. Of course, we must also discover whether relationships exist between what we know occurred during these periods and the specific nature of the deities ruling them.

Whatever the reader has noticed (and it would be a good idea to write it down for future reference), we continue with a discussion of the development of writing throughout the Great Cycle. The history of writing is not an arbitrary choice. The Maya saw writing as a gift of the gods, so it may be worthwhile to study what the different spiritual energies and the corresponding deities of the various baktuns may have had to do with its development. Writing is universally acknowledged to be a hallmark of human civilization; it is hard to imagine complex, diversified societies without it. It also deserves to be pointed out that as an intellectual activity, it engages the left brain. In studying the development of writing, however, we shall restrict the discussion to the odd-numbered Heavens, whose ruling deities, judging by their names and attributes, provide energies that help nurture the growth of the initial seed to a fruit. From figure 2.7, it is very apparent that the major steps facilitating written communications have been taken during odd-numbered Heavens.

The steps in the evolution of writing exemplify a new (but in fact very old!) way of looking at human history, in which it is only on a very superficial level that the emergence of various inventions are seen as the result of chance discoveries by individual geniuses. Instead, innovations in writing are understood to be reflections of spiritual energies nurturing creativity. Step by step the energies (or deities) of the seven

Heaven/ Growth Stage	Ruling Aztec Deity	Time Span	Development
Heaven 1 Sowing	**Xiuhtecuhtli,** god of fire and time	3115– 2721 BCE	First Sumerian logograms on clay tablets (3200 BCE); Egyptian hieroglyphs (3100 BCE)
Heaven 3 Germination	**Chalchiuhtlicue,** goddess of water	2326– 1932 BCE	Gradual development of Mesopotamian cuneiform
Heaven 5 Sprouting	**Tlacolteotl,** goddess of love and childbirth	1538– 1144 BCE	Consonant alphabetic writing in Canaan (1600 BCE); Chinese writing in the Shang Dynasty (1538 BCE)
Heaven 7 Proliferation	**Cinteotl,** god of maize and sustenance	749– 355 BCE	Complete alphabetic writing in Greece and Etruria (750 BCE)
Heaven 9 Budding	**Quetzalcoatl,** god of light	40– 434 CE	Papyrus codices in Rome (70 CE); Paper invented in China (105 CE); Mayan writing (250 CE)
Heaven 11 Flowering	**Yohualticitl,** goddess of birth	829– 1223 CE	First book, *The Diamond Sutra,* printed in China (868 CE)
Heaven 13 Fruition	**Ometeotl/Omecinatl,** Dual-Creator God	1617– 2011 CE	First daily newspapers in Holland (1618); First national mail service in Denmark (1624 CE)

Figure 2.7. Steps in the development of written communication during the odd-numbered Heavens of the Great Cycle

odd-numbered Heavens have supported the creativity of human beings and evolution in a certain direction.

The global development of writing, propelled by these odd-numbered Heavens, is an indication that the Long Count is a calendar for the historical evolution of the whole planet, based on the fact that the manifestations of its energy changes are global in character. Thus advances in the development of writing were made in several different locations, first in Egypt and Sumer, through several significant steps taken in China, to northern and northwestern Europe at its fruition. The civilizational evolution of seed to fruit nurtured by the Thirteen Heavens can thus be followed only in a global context.

What we may also see from figure 2.7 is that cosmic creation seems to be cyclical, in the sense that there is a cycle of human creativity. Close to the beginning of odd-numbered Heavens, significant innovations were made in the development of written communication. The even-numbered Heavens in between seem to represent periods when history rests. Activity and novelty emerging during the odd-numbered Heavens thus alternate with periods of rest during the even-numbered ones in a way that reminds us of an oscillation. This oscillation, in fact, is the same theme of death and rebirth that was so commonly expressed in ancient Mesoamerican myth. The Thirteen Heavens of the Great Cycle constitute a cyclical wave movement of history.

I caution against a narrow interpretation of the term *cyclical*. The wave movement of history is not cyclical in the sense of identical cycles being endlessly repeated (as is the case with many physical and astronomical cycles). The Mayan calendar describes evolutionary rather than strictly cyclical processes, so history is more like a spiral of evolution, in which similar types of events are favored at certain points in the cycle—for instance, at the beginning of odd-numbered Heavens. The results of these cyclically occurring bursts of creativity are never identical; a repetition of identical cycles does not generate evolution. History is rather a process resulting from stepwise increasing levels of consciousness, which are nurtured especially by the cosmic energies of the seven odd-numbered Heavens. The prophetic science of time of Native America and the Mayan calendar were developed to describe and understand this evolution of consciousness.

DAILY RESONANCE WITH COSMIC FREQUENCIES

Consider now that the common thirteen-day count, which is repeated twenty times in the Sacred Calendar of 260 days, is just a condensed version of the wave movement produced by the seven odd-numbered and six even-numbered Heavens of the much longer Great Cycle. The regular thirteen-day count of the tzolkin then becomes a kind of microscopic reflection of the energy changes that rule the long-term spiral evolution of human history. This explains why each day has its own spiritual energy, and why some days are said to be more favorable for creative activities than others. As can be seen in appendix B, this is exactly how the living Maya view the Sacred Calendar.

We can also draw an analogy with music. The common thirteen-day count (trecena) could be regarded as an overtone with a frequency 144,000 times higher than the basic tone of the Great Cycle (because there are 144,000 thirteen-day counts in the entire Great Cycle). The energy changes of the common thirteen-day count are then just as real as, but less powerful than, those of the Great Cycle. They are small ripples on the surface of the ocean of evolutionary change. And just as there are sounds of such a high frequency that not everyone can hear them, there may also be cycles of time with frequencies that are so high most modern human beings have difficulty noticing them. The thirteen-day cycle, codified as an integral part of the Sacred Calendar (20 × 13 days), is just such a high-frequency evolutionary cycle, which the Maya were aware of and which those of us who seek to integrate ourselves in the cosmic processes may want to retrieve. But how can we relearn? For the most part, our senses have been dulled by living in a society and environment that do not encourage our sensitivity to the divine flow of time. This dulling of the senses has been fortified by the almost universal rule of astronomically—that is, physically—based calendars. How then can we become more sensitive to, and align ourselves with, the flow of cosmic energies that help human beings evolve? Logically, the best way of getting into phase with cosmic creation is to follow the age-old classical Quiché-Maya-Aztec-Cherokee tzolkin count as a daily calendar, and then combine it with the tun calendar of cosmic wave movements. In that way we may *tune* in to divine creation and become part of its unfolding.

THE TUN: PILLAR OF PROPHETIC TIME

It may seem counterintuitive that prophetic cycles are nonphysical. After all, do not our moods change with the seasons? Are we not influenced by the full moon? Especially for those of us who live in higher latitudes, whether in the north or the south, it is almost a biological need to mark the solstices and equinoxes with festivities. Yet the seasonal cycles affect people very differently in different parts of the world. What in the Northern Hemisphere is the beginning of spring is the beginning of autumn in the Southern Hemisphere, and in the equatorial region it is difficult to talk about seasons at all. However it is divided, the seasonal cycle of the solar year does not produce synchronized global biospherical shifts. It is true that the full-moon cycle of 29.5 days affects everyone (probably because this is the natural duration of the female cycle in cultures that exist with no artificial light and hormones). Yet the important point is that endlessly repeated identical astronomical cycles could never explain the evolution of consciousness. In the next chapter we see that the evolution of consciousness is directly related to the heartbeat of Mother Earth, which the Mayan calendar helps us hear. This heartbeat comes from the inside, not the outside, of our planet. At most, astronomical cycles such as the solar year, the moon cycle, or the precessional cycle reflect endlessly repeated life-death cycles, such as summer-winter in certain parts of the world. Nonetheless, as we shall see, they have nothing to do with the divine plan helping human beings evolve to higher levels of consciousness.

Several ancient peoples in different parts of the world seem to have held many similar notions about prophetic periods. Both the Egyptians and the Maya, from very different perspectives, saw the time around the beginning of the Great Cycle as critical in the cosmic creation scheme. Did their respective calendars have similarities? Indeed, although there were some three thousand years between the times of their inceptions, the Egyptian and Mayan calendars had one striking point in common. The ancient Egyptians followed a calendar based on 360 days to which "the five days upon a year," the five days "when the gods were born," were added. During those five days, people were not supposed to show any feelings of joy or sorrow. Similarly, both the Maya and the Mexicas followed an agricultural calendar called the *haab* that consisted of 360 days plus five days when "the gods rested." These were five feared days when the humans could not count on the support

Figure 2.8. The shaman king 18 Rabbit (from the main plaza in Copán, Honduras) dedicating a new tun. Several such stone sculptures were erected by this king to mark the beginning of new prophetic time periods.

of the gods and ceremonies were performed to prepare for their return. It thus seems that both the Egyptians and the Mesoamericans wanted to express an awareness that the 360-day tun was more divine than the 365-day physical agricultural year.

In fact, in calendars used for religious and spiritual purposes by ancient peoples, the use of the 360-day period was the rule rather than the exception. The use of the 360-day year for spiritual purposes is known from ancient India, Peru, Scandinavia, Israel, and Rome. The Chinese used a religious calendar based on the 360-day year in parallel with the civil calendar of 365 days. In this difference between the 360-day religious year and the 365-day agricultural, or astronomical, year, we may see a parallel to the Maya and Mexica, who believed there were five days a year when the gods were absent.

A picture then emerges in which we may regard the 365-day period as a physical earth cycle, whereas the 360-day period is a spiritual cycle in the Universe of Holy Time. Although the spiritual meaning of the tun has largely been lost in today's world, we may retrieve much of it from a study of the Mayan calendar system as applied to human history. It is very important to note that in the prophetic context of the Mayan Long Count, one tun always follows another without any intermediary five-day period. For prophecy we need to follow a calendar that is based only on the spiritual cycles of creation itself and is not adapted to the physical particularities of our own planet. If prophecy is to be based on the spiritual reality, it must be linked to a calendar that describes this very reality.

There is much additional evidence that in ancient times the 360-day period called tun (meaning "stone," since the beginning of a new tun would be marked by moving a stone or erecting a stele) was considered among the Maya as a time of prophecy. We know, for instance, that at the time of the Spanish conquest, the Mayan priests would gather at the beginning of a new tun, rather than at the beginning of a new seasonal 365-day year, to deliver prophecies based on their calendrical books. In the few surviving prophetic books, called the Books of Chilam Balam, the Mayan prophecies were without exception developed for the tuns, or katuns (twenty tuns, or 19.7 years).

Because the Mayan interest in time was focused primarily on understanding the changing cosmic energies and how these influenced human life and cosmic creation in general, the use of tun-based time periods—for instance baktuns, katuns, and tuns—was the logical choice. The tun became the basic unit of the prophetic calendars of the Maya, and it plays the same role in the Long Count. If today we are to embrace a worldview in which consciousness is more important than matter, we too need to base our timekeeping on the nonphysical, invisible reality rather than on the physical. To grasp the distinction between the 360- and 365-day periods is a first and decisive step toward retrieving the ancient prophetic science of time.

3

...

The Cosmology of the World Tree

THE FOUR CARDINAL DIRECTIONS

If history is a cyclical wave movement of human creativity, how are its waves generated? To understand this we need to study the cosmology of the World Tree, which prevailed in Mesoamerica before the advent of the Europeans. When the Spanish first came to the land of the Maya in 1517 they were met by great painted crosses in Mayan centers of worship. At the Island of Cozumel off the Yucatán Peninsula, for instance, a pyramid had a three-meter-tall standing cross of lime in its courtyard. The Spanish learned that such crosses were representations of the World Tree, and that for this reason the crosses had been painted green (as, incidentally, Christian crosses in the area still are today). According to the Maya, the World Tree (fig. 3.1) is what creates and maintains the Four Directions of the world. After the Spanish conquest this World Tree was merged with the Catholic cross, but in the beliefs of the Maya

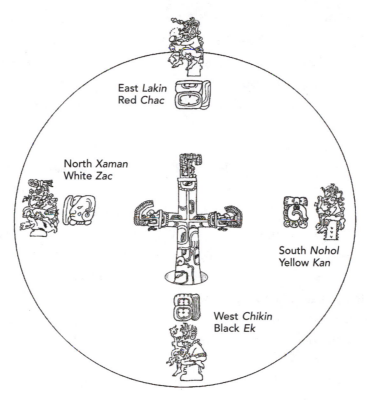

Figure 3.1. The World Tree of the Maya in the center of the Four Directions. Native American traditions commonly endow the geographical directions with different spiritual qualities (as in the Medicine Wheel). Among the Maya the Four Directions were symbolized by directional deities, so-called Bacabs, as seen in this illustration.

this new Christian cross still retains much of its ancient meaning as the center of the cosmos.

The view of the world as having four directions and four corners is common to the cosmologies of all Native American peoples, expressed, for instance, in the well-known Medicine Wheel used farther north on the continent. The Four Directions are associated with different energies and colors that give them different qualities. The Four Directions are related to the center of the world, called Yaxkin by the Maya. At first sight, this may not appear very different from the standard European view, but the meaning given to these directions is clearly different from what is common today. In the Native American view, the Four Directions embody spiritual qualities that influence human life differently depending on the reigning time cycle. The spiritual winds generated by the Four Directions, dominating different time cycles, are the very essence of the Mayan view, whereas in the typical European view the four directions are seen merely as a passive coordinate system applied to the earth.

Yet organizing the world according to four directions associated with different qualities may sound natural to all of us. Most of us recognize, at the very least, significant mental and spiritual differences between the directions of East and West. Whereas we see the East as having been dominated by collective structures and a meditative tendency, the West is seen as individualist, extroverted, and action-oriented. What is the background of these differences? Where is the line that separates East from West? Where, geographically speaking, is the center of the planetary Medicine Wheel?

A preliminary hypothesis regarding the line dividing East from West puts it at longitude 12° East, through central Europe and central Africa, the midline of the continental mass of the earth, stretching from the westernmost tip of Alaska to the easternmost tip of Siberia. But because we are not primarily interested in physical geography but in spiritual geography, we must verify the existence of such a hypothetical midline some other way. To identify the dividing line between East and West in a spiritual sense, we must instead use the Mayan calendar and study the directions in which the spiritual winds have been blowing as the energies of the Thirteen Heavens of the Great Cycle have shifted. We can then track historical events occurring at the baktun shifts of the Great Cycle in relation to a hypothetical midline of the planet running through longitude 12° East.

THE WINDS OF HISTORY

As mentioned earlier, the first patriarchal human civilizations with writing, monarchies, large-scale constructions, and religious beliefs in a

Creator God emerged at the beginning of the Great Cycle. This surge took place in the area called the Fertile Crescent, which stretched from Egypt through Israel and Syria to Mesopotamia. Throughout the first half of the Great Cycle, this area remained a spearhead of civilizational development.

It was not until about the beginning of the seventh of the Thirteen Heavens of the Great Cycle, which began to dominate in 749 B.C.E., that higher civilization developed at longitude 12° East. A sign of the civilizational awakening nurtured by this Seventh Heaven was that the Etruscans in northern Italy (see fig. 2.7, page 28) developed a system of writing in the latter half of the eighth century B.C.E. At about the same time, the first stable settlements on the Palatine were established, which was reflected in the mythical year of the founding of Rome in 753 B.C.E. Among the Mexicas this Seventh Heaven was sometimes said to be ruled by Tonacatecuhtli/Tonalaciuatl, the male/female duality that provided sustenance to human beings, and sometimes by Cinteotl, the god of maize and sustenance.

Figure 3.2. Europe during the beginning katun (749–729 B.C.E.) of Heaven Seven in the Great Cycle includes the emergence of Etruria as a historical culture and the mythical founding of Rome (753 B.C.E.).

After the establishment of a historical culture at the hypothetical midline, it becomes possible to follow historical winds at baktun shifts in relation to this line. Thus the next Heaven began in 355 B.C.E. with the Persian king Artaxerxes III Ochus moving west to reconquer Egypt and Asia Minor and subjugating the Athenians (fig. 3.3). A movement from the East toward the planetary midline thus initiated the Eighth Heaven.

Around the following shift, at the beginning of the Ninth Heaven in 40 C.E. (fig. 3.4), an expansionist policy dominated in the newly

Figure 3.3. Europe during the beginning katun (355–335 B.C.E.) of Heaven Eight in the Great Cycle includes military campaigns of the Persian king Artaxerxes III, signifying the movement from the East toward the planetary midline.

Figure 3.4. Europe during the beginning katun (40–60 C.E.) of Heaven Nine includes expansion of the Roman Empire through the conquests of present-day England and Wales, Morocco, Algeria, and Bulgaria. Movement occurs away from the planetary midline.

established Roman Empire, which in the following twenty-tun period grew to include today's England and Wales, Morocco, Algeria, and Bulgaria. This expansion from the midline incidentally began simultaneously with Paul the Apostle going out on his missionary journeys to spread the Christian faith to Asia Minor and Greece. According to the Mexicas, Quetzalcoatl, the god of light, ruled this Heaven.

In 434 C.E., the Tenth Heaven (fig. 3.5) began its rule. In this very year Attila became the ruler of the Huns, a nomadic people from central Asia, and the most powerful ruler of his day. When the new baktun began he attacked central Europe from the east and pushed the Germanic tribes to sack Rome, which led to the effective collapse of the Western Roman Empire just as this new Heaven came to dominate. We may already suspect that the shifting energies of the Thirteen Heavens have had a lot to do with the rise and fall of different civilizations. The period that in Europe followed the onslaught of the Huns has come to be known as the Dark Ages, because during the following few hundred years no center of higher civilization existed there. In the cosmology of the Mexicas, Tezcatlipoca, the god of darkness, ruled this Heaven.

Europe did not wake up from the Dark Ages until the beginning of the ninth century C.E. when, a few years into the Eleventh Heaven, the embryos of the modern nations of France and Germany emerged from the division of the empire of Charlemagne. The rule of this new Heaven also led to a shift northward in civilizational development, which resulted in the vitalization of a previously quiescent part of the world, Scandinavia. This is directly relevant to our tracking of movements to and from the midline that divided the East from the West. As this Heaven began, so did the journeys of the Vikings (fig. 3.6). After having been settled for several thousand years, peoples living in Scandinavia suddenly began going out on bold travels that would eventually lead them all the way to Greenland and America. As part of this outpouring, the ancestors of today's Norwegians and Danes raided the British Isles and the Danish King Knud conquered England. In the other

Figure 3.5. Europe during the beginning katun (434–454 C.E.) of Heaven Ten in the Great Cycle includes the invasion of the Huns under Attila and the collapse of the Western Roman Empire. Movement occurs from the east toward the planetary midline in this even-numbered Heaven.

Figure 3.6. Europe during the beginning katun (829–849 C.E.) of Heaven Eleven in the Great Cycle includes the raids of the Vikings toward the west and the east. Movement occurs away from the planetary midline in this odd-numbered Heaven.

direction, Swedish Vikings made their way into the Russian river system and reached Byzans in 839. Traditional historical science has never produced a satisfactory explanation for why the Vikings began to embark on these raids. The Mayan chronology sheds new light on this phenomenon, however; the Viking raids started just as a new odd-numbered Heaven came to rule. Thus at the beginning of Heaven Eleven, the winds of history seem to have driven people away from the planetary midline.

This awakening in the north, incidentally, took place simultaneously with the collapse of the classical Mayan sites in Chiapas and Guatemala, providing another example of how a civilization may disappear at a baktun shift. This is the famous "disappearance" of the Maya mentioned earlier, which, like the collapse of the Roman Empire, occurred at a shift of Heavens. The classical Mayan culture disappeared because the winds of history no longer were favorable to it. At the beginning of the new Heaven, the frontier of human history moved markedly north, in both Europe and Mesoamerica. In the time cosmology of the Mexicas, Yohualticitl, the goddess of birth, ruled this Eleventh Heaven, and in Europe a rebirth was certainly at hand. Also among the Maya around Chichén Itzá on the northern Yucatán Peninsula and the Toltecs of Tula, this was a period of rebirth, expressed through the prominent role that was given to the worship of Quetzalcoatl, the god of light.

A movement that from a traditional historical perspective may be even more difficult to understand than the sudden raids of the Vikings or the disappearance of the Maya is the Mongol storm that came to dominate the beginning of the next Heaven. At the beginning of the Twelfth Heaven, the Mongols—who, not so many years earlier, had been a small tribe herding sheep in the Gobi Desert—arrived in eastern Europe after having conquered all of China and Asia in between. These hordes continued their expansion until they arrived at the planetary midline (fig. 3.7). This even-numbered Heaven, which began with a violent movement from the East toward the midline, was ruled by Tlahuizcalpantecuhtli (good pronunciation exercise!), the god who ruled before dawn.

Figure 3.7. Europe during the beginning katun (1223–1243 C.E.) of Heaven Twelve in the Great Cycle. The Mongol storm hits eastern Europe from central Asia.

Here we should pause to consider two things.

First, in this progression we are using a chronology, that of the Great Cycle, which is ultimately based on the tun, the divine 360-day year. It was exactly at shifts between baktuns that these movements from the East hit Europe. In the same year the Tenth Heaven began, Attila became the ruler of the Huns; in the year 1223, when the Twelfth Heaven began, the Mongol storm reached the same area of eastern Europe. Had we used a calendar based on the physical year, the moments when these migrations from the east hit Europe would not have fit into a precise pattern. The winds of history, generated by the deities of the Four Directions, follow a baktun-based chronology.

Second, we should note the massive character of these violent movements. The empire created by Genghis Khan and his heirs around the beginning of the Twelfth Heaven was to set its mark on Russia and Asia for several centuries. In 1995 the *New York Times* elected Genghis Khan the Man of the (previous) Millennium, presumably not because of his good behavior but because of the impact the empire he created had on history. The Mongol storm created a unified field of the Eurasian continent, which allowed it to be crossed by a variety of phenomena from East Asia to Europe, such as gunpowder, book printing, the compass, and the plague. The movements coinciding with these baktun shifts are no minor events dug up just to prove a theory. The Mongol storm created the largest empire in human history, and the movements taking place at the beginnings of other baktuns had a similarly vast impact.

Finally, in this progression through the latter baktun shifts of the Great Cycle, we arrive at the rule of the Thirteenth Heaven (fig. 3.8). This Heaven was ruled by Ometeotl/Omecinatl, the Supreme Dual Creator deity who was seen as a dual god, both male and female, yang and yin combined, also called the god and goddess of duality by the Mexicas (fig. 3.9).

In central Europe, along longitude 12° East, the beginning of the rule of this Heaven was marked in 1618 by the Thirty Years' War between Catholics and Protestants, which established the survival and preeminence of the latter creed in northwestern Europe. In the north, as part of that scenario, Sweden temporarily became a major European power following its conquest of the Baltic coast of Russia in 1617 and the seizure of present-day Latvia and parts of Prussia from Poland. A small expansion westward was also evident in the founding of a colony in North America in 1638, and so the Thirteenth Heaven also began with movements away from the midline. Due to the intervention of Sweden and France in the second half of the Thirty Years' War, and the Dutch struggle for independence from Spain, the Holy German Roman Empire—which had been instituted by the pope and had dominated Europe during the previous two baktuns—lost its power. On the level of consciousness, this created an opening for a new mentality—that of the modern world—to be expressed.

The previous presentation has obviously been rather Eurocentric and so could create the erroneous impression that the Mayan calendar does not apply to other parts of the world. There are several reasons

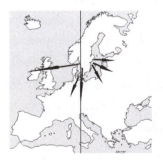

Figure 3.8. Europe during the beginning katun (1617–1637 C.E.) of Heaven Thirteen in the Great Cycle; includes the expansion of Sweden and the interventions of the Nordic countries in the Thirty Years' War.

Figure 3.9. The creator couple, Ometeotl and Omecinatl, the god and goddess of duality, rulers of Heaven Thirteen. This deity of two gods in one symbolizes the yin/yang polarity that was established in this Heaven.

that I have limited the discussion to Europe. The first is space. The second is to demonstrate the existence of the planetary midline, a projection of the trunk of the World Tree, which has made central Europe an area where the effects of the energy shifts of the Mayan calendar are especially sharp. Third, since the Mayan calendar system was not used in this part of the world, we know for certain that the movements we have observed are not the results of self-fulfilling prophecies. Instead, they are the objective results of the directional winds generated by the World Tree.

If we summarize the various movements from and toward the planetary midline at the baktun shifts of the Great Cycle, the very clear pattern of historical winds presented in figure 3.10 emerges. There we can see that Heavens with odd numbers initiate movements from the planetary midline toward the east and west, whereas Heavens with even numbers initiate movements from the east toward the midline.

The Mesoamerican notion that the various Heavens are ruled by deities that blow winds in four directions has thus now been directly verified. Judging from the pattern of violent migrations presented in figure 3.10, it also seems as if one divine force rules the seven Heavens

Heaven Number	Time Period	Movements from the Central Line of 12th Longitude	Movements from the East
7	749–729 BCE	Settling of Rome	
8	355–335 BCE		Persians go West
9	40–60 CE	Expansion of the Roman Empire	
10	434–454 CE		Huns go West
11	829–849 CE	Raids of the Vikings	
12	1223–1243 CE		Mongol storm
13	1617–1637 CE	Sweden	

Figure 3.10. Violent migratory movements away from and toward the planetary midline at the very beginning of each of the later Heavens of the Great Cycle

with odd numbers, and that this energy is somehow different from the energy that rules the six Heavens with even numbers.

In addition to these winds to and from the midline in the western and eastern directions, history seems with every new Heaven to have been blowing north, at least based on what we know from the Northern Hemisphere. The center of civilizational development moved from the southern locations of Egypt, Sumer, and Crete at the beginning of the Great Cycle to Greece and Rome during Heavens Seven and Nine and to Germany during Heaven Eleven. Typically, Denmark becomes a great power during Heaven Eleven, whereas Sweden, even farther to the north, becomes one at the very beginning of the Thirteenth Heaven. This northward movement of the frontier of European history had parallels in other parts of the world. In the Americas, innovative civilizations moved from Chiapas and Guatemala during Heavens Nine and Ten to Yucatán during Heaven Eleven, to central Mexico during Heaven Twelve, and to North America proper during the Thirteenth Heaven.

This pattern of the winds of history can be explained if we assume that a cross of invisible consciousness-field boundaries is introduced by the earth (fig. 3.11a, page 46) each time a new Heaven with an odd number begins its rule. This Invisible Cross is an expression of the World

Tree. According to the Maya, the First Father raised the World Tree, giving rise to the Four Directions, at the beginning of the first of the odd-numbered Heavens of the Great Cycle. Since the invisible World Tree generates powerful migratory movements throughout the Great Cycle, it makes sense that the Maya described the raising, or activation, of the World Tree by the First Father as the crucial initiatory event of creation.

The World Tree upholds a yin/yang polarity that introduces a creative tension along the lines where it oscillates, which in turn gives rise to movements in different geographical directions. The branches of the World Tree thus generate "winds," which is perfectly in line with Mesoamerican mythology. During the odd-numbered Heavens movements north (and probably also south) from the equator, as well as in both the eastern and the western directions away from the planetary midline, are generated. In contrast, when this cross disappears at the beginning of Heavens with even numbers, movements are directed toward the planetary midline.

Judging from this study, the arms of this cross, as hypothesized, go through longitude 12° East and the equator. The arm (trunk) that divides the planet into a Western and an Eastern Hemisphere goes through Rome and Copenhagen, whereas the arm (branch) that separates the Northern and Southern Hemispheres is the equator. If we accept that the creation field of the planet is fundamentally organized in this way, we must conclude that the center of the World Tree is located in central Africa, in an area of what is currently the nation of Gabon.

We may now understand why Ometeotl/Omecinatl was also called "Two Gods," because when he/she ruled the world it was divided into two halves, West and East, dominated by yang (light) and yin (darkness), respectively. The ancient Mesoamerican view has a much greater explanatory power than our current worldview. For our understanding of the past, the notions of the World Tree, the Four Directions, and the thirteen deities ruling the baktuns of the Great Cycle seem very meaningful indeed.

The history of humanity as it has developed through the Thirteen Heavens of the Great Cycle may thus be described as a wave pattern,

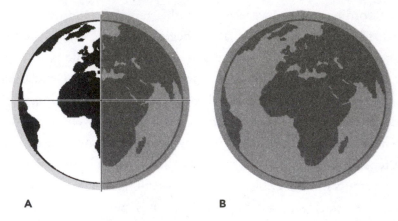

Figure 3.11. The global creation field during the rule of (a) odd-numbered and (b) even-numbered Heavens

and it is the World Tree, or Invisible Cross, that has served as the wave generator driving human history. It is thus not surprising that the World Tree—and the numbers 7 and 13—have had a central place in the mythologies of peoples all over the world. This World Tree creates wavelike alternations between dualist and unitary Heavens. The resulting two types of creation fields are shown in figure 3.11. The alternations between these two types of Heavens generate the cosmic drama that is enacted by the forces of light and darkness, male and female, West and East, and yang and yin. And, as mentioned, among the Mexicas the deities ruling the odd-numbered First, Seventh, and Thirteenth Heavens—Xiuhtecuhtli, Tonacatecuhtli, and Ometeotl, respectively—were all dual male/female creator gods.

The polarized creation field of Heaven Thirteen (fig. 3.11a), ruled by the Two Gods, meant the beginning in 1617 of a period of almost four hundred years of increasing world dominance by the West and a separation of the cultures of West and East. The buildup of the British Empire began with its first trading post in India in 1615 and the arrival of the Pilgrims in Massachusetts in 1620, the starting point of the development of the United States. Linked to this Western dominance are a number of phenomena that emerged around the beginning of the Thirteenth Heaven, such as Protestantism, capitalism, and science, expressions of the left hemisphere of the human brain. It is ultimately this duality, which was initially introduced when the First Father raised the World Tree in 3115 B.C.E., that has created the patriarchal, judgmental, evaluating mind. Although everyone in the modern world is in resonance with the global creation field of the Thirteenth Heaven, it has favored the West, because its light has preferentially shone on this hemisphere. The victories of the Protestant armies in the Thirty Years' War were a result of the establishment of this new creation field, and typically, the scientific revolution, initiated as the Thirteenth Heaven began by individuals such as Kepler in Prague and Galileo in Florence, was conceived exactly under the midline. Later, as might be expected, the center of gravitation of this mind-set gradually moved west through the France of Descartes, Pascal, and Fermat and Isaac Newton's England, to end up in the United States toward the latter half of the twentieth century.

At the beginning of the Thirteenth Heaven the French scientist and philosopher René Descartes formulated its new philosophy, which is sometimes referred to as the Cartesian split. Descartes argued that material and spiritual were two separate realms of existence, and in recent decades Descartes has been more or less personally blamed for the separation between human beings and nature, or between body and mind. But Descartes did not create this split, which for many generations was taken for granted. The World Tree created the split. Descartes simply composed the most eloquent formulation of a philosophy that described the split in perception imposed on human beings by the Thirteenth Heaven. Only recently has the downside of this perception become evident to a broader group of people, and we will later understand why.

The Invisible Cross and its polarizing effects on the consciousness field of the planet have also created an inner sense of dissatisfaction in the modern mind, an experience of not being whole. Reminiscences of an earlier wholeness and myths about a Fall are almost universal in ancient cultures. The raising of the World Tree by the First Father at the beginning of the Great Cycle amounted to exactly such a Fall.

Concomitantly, a belief emerged in a punishing God with whom humans were to some extent in conflict, which is fairly evident in the Hebrew Scriptures. All the major monotheist religions that developed during the course of the Great Cycle were profoundly influenced by its dualist consciousness and so have tended to look upon the world through a good/evil, right/wrong dichotomy generated by its yin/yang polarity. Because of this, these religions are all to varying degrees uncompromising and have tended to engage in sometimes violent conflicts about right and wrong. For the same reason, religions generated by this consciousness field have created a hierarchical structure among their followers, often in direct contrast to their professed egalitarian message.

We may then understand the modern mind, created through holographic resonance with the Thirteenth Heaven, as the fruition of a process that began with the Fall. The ancient myths about a Fall—the expulsion of Adam and Eve from the Garden of Eden in the Bible, the shooting of Seven-Macaw in the Popol Vuh, Seth and Osiris in Egypt, and Ask and Embla among the Norse (most with prominent roles for the World Tree)—are thus metaphors for a process that actually took place on the level of consciousness. The result of this Fall is the "halfness" of the modern mind, a mind that rarely has the peace to just be and instead is always looking for ways to change and become something different. Half of human experience, those qualities mediated through the Eastern Hemisphere, was in fact disfavored and filtered out by the Great Cycle.

The idea of such a drama, involving the male and female forces of the cosmos and recognizing both separation and unity between the two, is shared by many ancient philosophies. The Chinese notion of yang and yin may be the best known, but the adventures of the Mayan Hero Twins, Hunahpu and Xbalanque, and the corresponding twins of the Toltecs and Aztecs, Quetzalcoatl and Tezcatlipoca, are all metaphors for the same polarity. The fundamental truth that the consciousness field of the earth had the basic structure shown in figure 3.11a seems to have been known at some level by all ancient peoples. We can see this expressed, for example, in the sun wheel symbol for the earth used in Babylonian astrology (fig. 3.12), the Celtic cross, or the Medicine Wheel in North America.

Despite the universal symbolism of this polarity and the many parallels it is possible to draw between different mythologies, we should note that the Mayan chronology has something unique to offer that puts the world in a totally new perspective. This is the connection to the real world and to real history. No longer is the yin/yang cosmology merely an abstract philosophical principle. Instead, because of the exact determinations of the time periods that condition the pulsating drama generated by the World Tree, it is now evident that the yin/yang cosmology has a real foundation. This foundation is provided by wave patterns of both light and duality, which, as we saw in the previous chapter, nurture a spiral evolution of human creativity. This is very useful for understanding human history and our own place in it. All that is needed to understand the wavelike process of history—and the ensuing play of yin

Figure 3.12. The Cross of Vuotan, or sun wheel (from a Bronze Age tomb in Kivik, Sweden)

and yang—is to accept the existence of a World Tree serving as an all-encompassing oscillator operating according to the rhythm given by the Mayan chronology, the Long Count. Why not? All wavelike movements and vibrations need a wave generator, after all. The fundamental truth about the Mayan calendar is that it describes the exact timing of the pulses and vibrations emanating from the World Tree that determine the evolution of human consciousness.

As this wave pattern of history becomes visible via the Mayan chronology, we are presented with a choice. If we continue to use astronomical calendars we will probably continue to see history as a series of chaotic, random events. If, on the other hand, we choose to apply the Mayan calendar, we will become aware of the patterns and contours of a cosmic plan.

In fact, we may now see a real basis for prophecy in the winds generated by the World Tree. Hence it becomes entirely understandable that the arrival of the Europeans and Christianity had been prophesied in Mesoamerica. This arrival was only logical given what we now know about the winds generated by the World Tree, and the same can be said about the ensuing European colonization of much of the world. The course of events and exact physical manifestations may not have been predictable in detail. Yet the big picture had already been codified in the Heavens, and so those living in the Holy Universe of Time of the Mesoamerican calendars were able to prophesy what was to come.

THE HERMETIC PRINCIPLE AND THE GLOBAL BRAIN

The model I have suggested based on the Four Directions—with a midline separating two global hemispheres—is consistent with the so-called Hermetic principle: as above, so below. The qualities of the Four Directions point to a number of parallels between the "global brain" of the planet and the human brain. Our two brain hemispheres, right and left, have different functions because they are microcosmic reflections of the Eastern and Western Hemispheres of the planet (fig. 3.13).

The rational, action-oriented Western Hemisphere is a parallel to

Global Brain	Human Brain
Western Hemisphere	**Left hemisphere** (analysis and logic)
Eastern Hemisphere	**Right hemisphere** (synthesis and intuition)
Germany	**Hypothalamus** (central coordination)
Italy	**Hypophysis** (central coordination)
Nordic countries	**Epiphysis** (light monitoring)
Central Africa	**Cerebellum** (voluntary movement and emotions)
Hawaiian Islands	**Eyes** (vision)

Figure 3.13. Suggested parallels between the global brain and the human brain

the left, "male" brain hemisphere with its analytical thinking and centers for speech, calculation, and sequential logic. The meditative Eastern Hemisphere, on the other hand, parallels the right, intuitive, wholeness-oriented, "female" brain hemisphere, which is also the center for our spatial and artistic abilities. What this means is that the different mentalities we find typical for people from the East and West have ultimately been generated by the World Tree.

In this global brain structure, divided approximately along the trunk of the World Tree, the region of Germany-Italy corresponds to the hypothalamus-hypophysis complex, the regulatory center of the mammalian brain. In today's world, Italy and Germany may not often be thought of as being in such central positions, but during the Renaissance and the Reformation they certainly were. Through the Roman and German-Roman Empires, as well as the papacy, this region dominated the history of Europe for some two thousand years. Farther north along the same midline, the Nordic countries, and Sweden in particular, represent the pineal gland or epiphysis. Central Africa, the origin of humankind, corresponds to the cerebellum, the oldest part of the brain, with its critical role in balance, motor skills, and emotionality. On the opposite side of the world, the Hawaiian Islands, with its international assembly of telescopes on the rim of the volcano Mauna Kea and the giant telescopes of Maui, correspond to the eyes through which this planetary brain looks out into the cosmos.

Also, just as people normally use the left hemisphere of the brain for their external interactions, the world has, at least until now, looked mostly to the West, and especially to the United States, for leadership. While most peoples in the Eastern Hemisphere live in a more historical context, North Americans typically live in the present and have a more operational attitude toward life. With the framework of the global brain, the different mentalities in different areas of the world can easily be seen as resulting from a process of creation that generates different yin/yang fields in the planetary context. Depending on where we live, our individual consciousness is conditioned by the mentalities of the Four Directions. According to the Maya, the World Tree was the source of all life, human beings included, and so it has turned this earth into a living, pulsating world where all parts are connected and interdependent. (See fig. 3.14.)

In this perspective it is not surprising that the scientific revolution began as the dualist Thirteenth Heaven began its rule. This new Heaven meant that Light would fall not only on the Western Hemisphere, but also, through holographic resonance, on the left-brain hemispheres of human beings. As a result, the functions normally associated with the left hemisphere—analysis, mathematical calculations, sequential logic, and so on—were highlighted as the Thirteenth Heaven began its rule. These are exactly the functions that were prerequisites for the development of the scientific mind and the revolution in thinking that this has brought about. Kepler's three laws of planetary motion, presented in *De Harmonice Mundi* in 1619, were groundbreaking. These were the first mathematical laws of nature.

Figure 3.14. The sarcophagus lid from the tomb in the Temple of the Inscriptions in Palenque—one of the most famous of all Mayan pieces of art. The Great Pacal is seen here sinking into the roots of the World Tree, later to be reborn at its branches.

It is important to point out that in the cosmology of the four directions, all the different parts of the global brain are needed, and each has a role to play in the integrated activity and balance of the larger whole. To draw another parallel with the individual human brain, new creative thinking typically arises in an interaction between the two brain hemispheres, in which the right hemisphere generates the new ideas and the left hemisphere evaluates them. Unfortunately, however, the two hemispheres of our planet do not interact in this way. While the technology of the West has certainly been exported to the East, and much philosophy in the past few decades has gone from the East to the West, the two modes of thinking are as yet not integrated; they merely coexist side by side. Fruitful creative interaction is hampered by the lack of a common framework, a recognition that the two modes of thought are generated by the same source. The cosmology of the World Tree and the calendar of ancient Mesoamerica could help such a holistic framework emerge. The thinking of modern humans is out of balance partly because it has shut out the original West.

THE PHYSICAL BIOLOGY OF HOLOGRAPHIC RESONANCE

How are these yin/yang fields of the global brain generated, and what is their nature? Why is reality perceived differently in the East and the West, and why is creativity expressed in such different ways in the two hemispheres of the planet? Although these yin/yang fields are not among the commonly recognized fields of physics, a discussion of holographic resonance in terms of physical biology may be valuable to grasp the mechanism by which information from the global creation fields is expressed in the thinking of human beings.

An interesting key to holographic resonance is the fact that the human brain in a relaxed or meditative state displays alpha waves with a frequency of 7 to 13 Hz (cycles per second), and that the lower end of this is close to the so-called Schumann resonance of 7.8 Hz that envelops the earth. The Schumann resonance is a standing wave of electricity produced in the ionospheric cavity between the earth and the ionosphere, which is rich in electrical discharges due to thunderstorms and so on. The Schumann resonance is a constant of this particular planet and does not change over time, except for a minor variation of ± 0.5 Hz due to varying solar activity (contrary to rumors concocted in the New Age community that it is increasing). The reason for the stability of the Schumann resonance is that its value depends on just two simple figures: the speed of light at 186,000 miles per second (300,000 km/s) and the circumference of the earth of 24,450 miles (40,000 km). Hence it has a value close to 186,000/25,000 = 7.5 cycles/s = 7.5 Hz.

Because the circumference of the earth's surface depends directly on its radius (r = 24,450/2π = 3,893 miles), the lower end of the alpha frequency in human as well as other mammalian brains has the same frequency as an electrical field produced by a standing wave circulating with that radius. Thus, regardless of what meaning, if any, we would attribute to the mental resonance with thunderstorms measured at the Schumann frequency, it seems that the human mind is in a relaxed state when it is in resonance with a field at the surface of the earth. Since this surface is where we live, this is probably not a coincidence; and in this sense we may look upon the meditative alpha state as a means of coming home.

Now, of course, if there is a simple inverse proportionality between the radius of a sphere and the frequency of a standing electromagnetic wave enveloping that sphere (see fig. 3.15), it becomes interesting to

Sphere	Radius from Earth's Center	Frequency
Magnetopause/plasma sheet	60,000 km	0.8 Hz
Outer Van Allen Belt	25,000 km	2 Hz
Inner Van Allen Belt	12,000 km	4 Hz
Earth's crust/mantle	6,370/6,360 km	7.5 Hz
Outer core	3,500 km	13.5 Hz
Inner core	1,200 km	40 Hz

Figure 3.15. Radiuses of spheres surrounding the center of the earth, along with the corresponding frequencies of standing circular electromagnetic waves

investigate other spheres around the earth's center (see fig; 3.16). A few spheres where an electric potential creates standing electrical currents are listed in the table below with the corresponding frequencies of these waves.

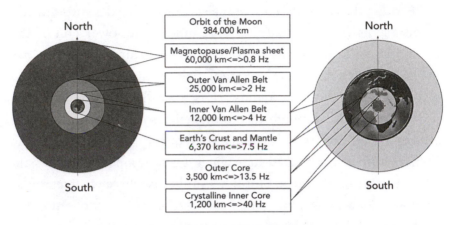

Figure 3.16. Distances of different spherical fields from the center of the earth, along with their corresponding frequencies

What we have is a series of spheres, of both a geological and an atmospheric nature, that all share the earth's midpoint as their center. These spheres are probably the most significant boundaries between the different concentric segments that constitute the earth. In more esoteric language, they could probably be called ethereal bodies. Figures 3.17 and 3.18 compare the frequencies of waves at these boundary surfaces with what is known about the different types of electromagnetic waves of the human brain.

The concordance between these traditional ranges for the frequencies of different types of brain waves and the radiuses of the earth's spheres is remarkable. This concordance, moreover, is not something we need to go to obscure scientific journals to verify. It is among the most basic knowledge for students of the respective fields; and for a scientist used to experimental difficulties, the concordance is astounding. One is tempted to draw the conclusion that the different mental states

Wave Type	Mental State	Frequency
Delta	Deep sleep	1–4 Hz
Theta	Light sleep; drowsiness	4–7 Hz
Alpha	Relaxation; meditation	8–13 Hz
Beta	Mental concentration	13–40 Hz

Figure 3.17. The main types of brain waves with their corresponding mental states and frequencies

of the human being are simply functions of resonances with different spherical layers surrounding the earth's center. When, for instance, we are in the alpha state, we are in resonance with a field of consciousness in the earth's mantle (extending from the earth's crust to the outer core), and when we are in deep sleep we are in resonance with the earth's outermost resonance body. I do not mean to suggest that the resonance itself is electromagnetic in nature, but rather that when we are in resonance with, or when we *tune* in to, these different mental earthly fields, electromagnetic currents with the same frequencies are activated in our brains.

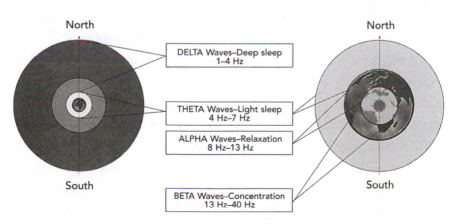

Figure 3.18. The physical and nonphysical spherical segments corresponding to different states of mental awareness

The frequencies of the various brain waves discovered in humans seem to be shared by most organisms, so this finding may not, after all, be that surprising. If these brain waves had their origin only in the internal functioning of the organism, and thus were unrelated to external fields, one would have expected a considerably greater variation between species in the frequency of brain waves than what is observed. The remarkable correspondence above is thus strongly suggestive of a resonant relationship between different mental states and different spheres of the earth. Such a relationship would provide proof that human thinking, and mental activities generally, do not go on inside of the individual, but emerge in an interaction with consciousness fields of the earth and of the cosmos.

THE SCIENCE OF CONSCIOUSNESS OR GEOLOGICAL PSYCHOLOGY

The state of sleep, either deep or light, delta or theta, means being in resonance with the atmospheric mental layers, whereas the awake state means resonance with the solid earth. From this perspective it is not surprising that as we sleep we go deep out in space, since in these layers there is the least grounded, structured control of our thoughts or visions. In a dream state, connections may be established that we cannot always understand by applying sequential logic. As mentioned, it also seems reasonable that the relaxed alpha state corresponds to the biosphere.

Perhaps the most interesting aspect of this relationship, however, is that the beta-type brainwave frequency arises from holographic resonance with the earth's inner core. How does this fit into the perspective given by the Mayan calendar as presented in this book? First, the ancient peoples of the earth not only described a World Tree, but also a World Mountain. The World Tree is the polar axis, projecting perpendicular planes (branches) onto the surface of the earth, but it is anchored in the World Mountain in the center of the earth. The planetary midline is only the most important of these branches of the World Tree, creating the alternating yin/yang polarities that propel the evolution of consciousness. Although the World Mountain may be less commonly used as a metaphor than the World Tree, it is said that the pyramids and many mounds were intended to symbolize it. And if the Mayan pyramids were meant to symbolize the Underworlds, perhaps there would also be a link between these and the earth's core. What would such a link serve to explain?

It is difficult to elaborate this link because we have very little certainty about the nature of the inner core of the earth—the World Mountain. It has recently been estimated to be as hot as the surface of the sun. This immensely dense core of iron, possibly mixed with other elements, is under a pressure of about 10 million tons per square foot (1 million tons per square meter). Scientists working at Harvard have discovered an anisotropy (irregularity) in its shape, meaning that this core is not a perfect sphere, but it has at least one line of symmetry. The researchers have concluded that the immense pressure to which the iron is subjected at the earth's center (imagine the weight of a million cars on top of you) has given it a crystalline structure.

Although we do not know the exact nature of this crystalline structure, the fact that it does exist is important. Given that the ancients, with their sometimes highly developed intuition, built pyramids as symbols of the World Mountain (and possibly to create enhanced resonance with it), we may speculate that it has an octahedral structure. By outward projection, such a structure, with perpendicular corners, could create the planetary midline and the corresponding yin/yang fields of the global brain, upon which much of the discussion in this book is built. Rather than a real octahedron, however, I think a pseudo-octahedral structure is more probable, more like a cross between a pyramid and a stupa (which also is a symbol of the World Mountain with a World Axis at its top), resulting in a multifaceted surface. This may also be why the ancient Mesoamericans labeled the major creation cycles "Underworlds," as we discuss later. These Underworlds, as the terracelike pyramidal models of the World Mountain seem to imply, might originate in the crystalline structure of the earth's inner core, where different layers are activated according to the preset pattern of the Mayan calendar. Could the Nine Underworlds correspond to nine sequentially activated layers of iron crystals in the earth's inner core?

The yin/yang polarities of creation would then originate at the different sides of this basically octahedral crystalline structure and radiate outward to reach and influence the people at the crust. Through the underlying synchronic order of the ether, this crystalline core would receive information about polarity shifts from the galactic core, which in turn would have received it from a hypothetical Universal Central Sun beyond the horizon of time in our visible universe. Although changes in yin/yang fields would occur in synchrony throughout the universe, the earth's inner core would serve to anchor and radiate them.

I am speculating freely here. All we know for sure is that the earth's inner iron core is crystalline and has one symmetry line. But if it has one such line, why not many? The crystalline state implies symmetry lines. At the immense pressure and temperature of the earth's inner core, matter probably displays properties that are completely unknown from our everyday experience at its crust.

Also of interest in this connection is that according to dowsers, the surface of the earth is covered by grids of straight lines—called Curry and Hartmann lines—that are not related to the common physical fields of electromagnetism and gravitation. Since the only objects that exhibit straight lines in nature are crystals, it is not unreasonable to speculate that these grid lines are projections of a consciousness grid from the earth's crystalline inner core. Those who are familiar with the energies of such grids and their crossings, with male and female healing points and so forth, can attest to their strong influence on consciousness.

Now, if we accept that the human brain is a microcosm of the global brain, it would only be logical that the brains of human beings are also covered by a microscopic grid. This grid would constitute our consciousness, which we ultimately owe to the earth and the cosmos. In the awake beta state, our brains would be in resonance with the earth's inner core. Beta waves of 40 Hz, according to experts in the field, are closely linked to perception and consciousness. It can hardly be an accident that the most highly structured and focused activities of the human mind at the beta level would be in resonance with the earth's inner crystalline core, which is the earth's most structured part. Beta-level thinking is the most logical, structured, and sequentially ordered. (Note that we sometimes use the term *crystal clear* to describe logical arguments!) The idea that humans channel information from the earth now loses much of its mysticism. How would it be possible not to channel the earth? We all do it. A person's creativity may depend only on his or her ability to be in resonance with different earthly spheres at the appropriate time.

The consequences of these ideas would easily take up a book of their own, and here I will only hint at how the consciousness grid may explain the different mentalities of East and West. Richard Nisbett has demonstrated in *The Geography of Thought* that the differences in thinking between Easterners and Westerners are based not only on differing philosophical traditions, but on differences in their very psychology and perception of the world. Where Westerners see objects with attributes (trees), Easterners see larger contexts and relationships (forests). Could the atomizing Western mind-set, which was behind the emergence of science as the Thirteenth Heaven began its rule in 1617, be a function of the grid of consciousness becoming activated as the light began to shine on the Western Hemisphere in the yin/yang duality of this Heaven? Could the invention of calculus have been related to the new grid, as well as the much more exact subdivision of time based on the pendulum clock? Would it be easier to order things in time and apply sequential logic if your brain had a more precise built-in grid for time? Could the invention in the West of the phonetic alphabet, which divides sounds linearly in time, be a function of a more powerful inner grid becoming evident during odd-numbered Heavens? Is it an accident that early European Baroque paintings from around the time of the beginning of the Thirteenth Heaven are among the most photographic ever to have been made, or is it related to the precision of an inner grid defining human perception? If the crystalline core of the earth generates a resonant grid of consciousness, an enormous number of phenomena related to the differences between brain hemispheres and global hemispheres would be explained.

If concentrated mental activity arises from resonance with the earth's inner core, this may also shed new light on the rise in sensitivity to electricity in the past decade. As our civilization has become increasingly electrified, the widespread use of alternating current (50 Hz in Europe and 60 Hz in the United States) may be increasingly interfering with people's natural resonance with the core of the earth, the World Mountain. (There is also a set of brain waves labeled gamma, ranging from 30 to 90 Hz, that may be even more disturbed.) At the present awakening, these alternating currents may in some people conflict with their evolution of consciousness, resulting in adverse mental conditions that also generate physical symptoms.

The relationship described here between brain frequencies and different segments of the earth probably has consequences for meditation, because enlightenment, or awakening, is a matter of raising the frequencies of the mind. The dramatic rise in interest in crystals over the past two decades is likely also related to this relationship. Today as we approach the Universal Underworld (more on this in chapter 9), raising our frequencies and awakening our minds through resonance with the earth's inner core are tantamount to following a path toward enlightenment. Increasing our frequencies also means resonating more deeply with Mother Earth—resonating with her very core. If the evolution of consciousness is about resonance with the earth's inner core, it should be clear that astronomical calendars lack prophetic relevance.

4

• • • •

God and the Historical Religions

THE GREATEST ENIGMA

Does God, a universal intelligence behind all things, exist? This is one of the most commonly asked existential questions, and certainly the answer has very significant consequences when it comes to the meaning of life and what we are to make of it. Can the Mayan calendar help us find the answer? If we accept the characterization of the Mayan religion by archaeologists and anthropologists as polytheistic, this would seem unlikely. I look upon the various deities participating in the Mesoamerican creation stories, however, as different manifestations of the One Intelligence. Although the ancients are not here for us to ask, it is wise to be open to the possibility that at least some of them held such a view.

Much of the discussion regarding the existence of God is incidentally colored by the fact that this universal intelligence of the cosmos has usually been associated with the portrayal of

God offered by a specific historical religion and its holy scriptures. In the Western world, for instance, God is typically seen through the Christian filter and linked to the particular forms of worship prescribed by this tradition. Especially with the Thirteenth Heaven, this view of God has become very abstract, and devotion to the living cosmos, including various spiritual forces, has been condemned as worship of idols, or polytheism.

From our knowledge of the Mayan calendar we may now begin to realize, however, that the consciousness grid through which we experience divine reality undergoes an evolution. "God" looks different in an era favoring the left-brain hemisphere from how "God" appears in an era that does not. The World Tree has been the most powerful and decisive influence on the evolution of religions, or at least has generated the initial impulse propelling their growth. Any historical religion gives us only one particular view, seen through a particular frame of consciousness generated by the specific Heaven that ruled as it was conceived. It is thus meaningful to explore the nature of God from the more universal perspective provided by the Mayan calendar, which highlights the nature of the filters in the human perception of the divine.

By studying how the wave movement of human consciousness has influenced human religiosity, and later the absence of it, we may learn something about the greatest enigma, the existence of God, from the Mayan calendar. We need to realize that where people have been living in relation to the Invisible Cross has profoundly influenced their perception of divine reality. Thus there are distinct differences between the traditional religions that originated in the East and those that originated in the West—and in between.

THE CHALDEAN-JEWISH-CHRISTIAN TRADITION

Our study of how the Thirteen Heavens have influenced the evolution of religions begins with the tradition that has attracted the largest number of professed adherents. This is the so-called Chaldean-Jewish-Christian tradition (Chaldea is the Biblical name for Sumer), which, during the course of the Thirteenth Heaven, has come to dominate not only all the Western Hemisphere but other parts of the world as well. This tradition has been driven forward by a series of pulses that has each developed its own particular expression. In tracking the historical evolution of this tradition, which parallels the development of writing, we again concentrate on the odd-numbered Heavens, since these are the ones markedly influenced by a creative yin/yang polarity. In figure 4.1 this tradition has been assigned to the center, around the East-West midline through Europe, although its origin is to be found somewhat east of that.

It is difficult to know anything for certain about what religious beliefs people had before the beginning of the Great Cycle, because there are no historical records from that time. As far as we know, however, the notion of an omnipotent Creator God first emerged during the First Heaven of the Great Cycle. Already at this early time the Sumerians seem to have worshipped an omnipotent God of Heaven who went by the name of An. In ancient Sumer, however, this Creator God shared his place with a plethora of harvest gods, house gods, and city gods, as well as many others.

From the biblical patriarchs, who during the Third Heaven brought the belief in this Creator God from Chaldea to Canaan, we learn that the extended pantheon of the Sumerians had already lost much of its previous role. Yet it was only with Moses (in resonance with Heaven Five), and the Ten Commandments that Moses brought to the Jews as a message from God, that a consistent monotheist creed emerged. The first of these Ten Commandments is "You shall not have other gods besides me" (New American Bible). Although this commandment expressed a pure monotheist creed, it was limited in that it seemed to apply only to the Jewish people. God was then seen as the God of the Jews, and at the time it was hardly a concern of the Jews whether other peoples would believe in him. *You* in the First Commandment thus refers to the Jews, who were seen as a people chosen by God.

The idea of God being exclusively the God of the Jews did not change until at the beginning of the Seventh Heaven. Then the great

Heaven No. Growth Stage Time Span	West/America	Center/Europe	East/Asia
Heaven 1 **Sowing** 3115–2721 BCE		Sumer's An	
Heaven 2 2721–2326 BCE			
Heaven 3 **Germination** 2626–1932 BCE		Abraham's move to Canaan (2300 BCE)	
Heaven 4 1932–1538 BCE			
Heaven 5 **Sprouting** 1538–1144 BCE		Moses (1480 BCE)	Traditional Chinese (Shang); Vedic tradition
Heaven 6 1144–749 BCE			
Heaven 7 **Proliferation** 749–355 BCE	Zapotec Tzolkin (c. 550 BCE)	Isaiah (748 BCE); Zoroaster (c. 550 BCE); Deutero-Isaiah (c. 550 BCE); Pythagoras (c. 550 BCE)	Lao-tzu; Buddha (552 BCE); Reincarnation in India; Confucius (551 BCE)
Heaven 8 355 BCE–40 CE			
Heaven 9 **Budding** 40–434 CE	Quetzalcoatl in Teotihuacán	Jesus/Paul (33–37 CE); Christianity Talmudic Judaism	Buddhism in China (60 CE)
Heaven 10 434–829 CE			Islam (632 CE)
Heaven 11 **Flowering** 829–1223 CE	Second Quetzalcoatl in Chichén Itzá and Tula	Expansion of Christianity to northern and eastern Europe; Crusades; Height of papal power	
Heaven 12 1223–1617 CE			Second wave of Islam
Heaven 13 **Fruition** 1617–2011 CE		Expansion of Christianity; English Pilgrims (1620)	

Figure 4.1. Evolution of the historical religions, especially during the odd-numbered Heavens of the Great Cycle (see also fig. 4.6, page 81)

Jewish prophets—Isaiah, Amos, and Hosea—began to teach; the Prophet Isaiah was called to his mission in the very first year (748 B.C.E.) of this baktun. Isaiah, who in the Jewish tradition is considered as the foremost of the prophets next to Moses, and who plays a similar role in the Christian tradition, taught that God was using invading peoples, such as the Egyptians and the Assyrians, to punish the Jews for their ungodly ways. This meant that the God of the Jews came to be seen as ruling other peoples as well, and even as determining the course of world history. In the second part of the Book of Isaiah—believed to have been written by another prophet, usually referred to as Deutero-Isaiah (the second Isaiah)—this change in thinking was fully developed. Deutero-Isaiah lived around 550 B.C.E., at the very midpoint of the Seventh Heaven. He taught that although the Jews were special in the eyes of God, he was nevertheless the God of all human beings.

This insight—that there is but one God who is the God of all humanity—may be the most fundamental religious insight of the Great Cycle, the very core of the spiritual understanding that it carries. Although the Chaldean-Jewish-Christian tradition continued to undergo change as the Great Cycle progressed through its later Heavens, this remained its most fundamental truth. And we should note that this belief in a universal God was first expressed at the midpoint of the Seventh Heaven, which is also the midpoint of the entire Great Cycle of the Maya.

Let us recall (see fig. 2.2, page 19) that the Seventh Heaven was represented by a butterfly rather than by a bird. This symbol of metamorphosis and transformation is particularly apt at this stage at the top of the seven-storied pyramid. In the Seventh Heaven the new frame of consciousness truly began to dominate, and at this stage the previous view of an enchanted cosmos with a multitude of spirits and deities metamorphosed into a belief in a Universal God. In the previous chapter we also saw that during the Seventh Heaven the creative tension of the World Tree began to manifest in a clearly evident way, generating winds that would spread beliefs across the planet.

What happened in the development of the Chaldean-Jewish-

Christian tradition at the beginning of the Ninth Heaven also seems fairly clear. Christianity split off from Judaism (the duality created by the World Tree often leads to separations). Thus Jesus is believed to have taught during the years 30–33 C.E. and to have been crucified in the last of these. Paul, who in many ways became the founder and organizer of the Christian religion, was converted in the year 37 C.E. He went on his first missionary journeys around the year 40 C.E. The decisive break between Christianity and Judaism seems to have been made at the Apostolic Meeting in Jerusalem in 49 C.E., only a few years into the Ninth Heaven. There the apostles decided that non-ethnic Jews, who did not abide by the Law of Moses in all its detail, could be part of the Christian community. Taking this step effectively led to a break with Judaism, a religion in which the Mosaic Law plays a predominant role, and so from this point Christianity started to spread as a distinct religion.

Differences between the Christian and Jewish faiths also contributed to the split. The Christians, emerging with the mentality of the new Heaven, emphasized forgiveness, in contrast to the ancient idea of "an eye for an eye" practiced not only by the Jews but everywhere else as well in earlier days. Jesus Christ also taught an end to all forms of sacrifice. Moreover, in the Christian view at the end of time there would be a Kingdom of God in which humans would have eternal life, and this kingdom was where the human soul was meant to go. The idea that life on earth has a purpose and that history is going somewhere had not previously been stated anywhere as clearly as it was in the New Testament.

Other than this, the Chaldean-Jewish-Christian tradition became universalist not only in theory, but also in practice, as a result of the strong winds generated by the Ninth Heaven. Jesus said: "Go ye therefore, and teach all nations." Christianity became a proselytizing religion, which aspired to be embraced by the whole world. As a step in this direction, a parish was formed in Rome, on the planetary midline, already in the fourth decade C.E., and there the institution of the papacy would later emerge. Throughout the new Heaven the following of the Christian faith and the political power of its church would continue to rise. Toward the end of this Heaven, Christianity became the state religion of the Roman Empire, and in some ways the Ninth Heaven may be looked upon as a breakthrough to a new light.

During the Tenth Heaven, the dark ages ruled by Tezcatlipoca, however, the leading role of the papacy was challenged both by the Irish and Byzantine Churches, and above all by Islam. As this Heaven came to a close, the importance of the papacy, located under the midline of the cross, was firmly reasserted as a function of its initiative to reinstitute the Holy Roman Empire. During the course of the Eleventh Heaven, 829–1223 C.E., many peoples in northern and eastern Europe were converted to the Christian faith.

Finally, as already mentioned, at the beginning of the Thirteenth Heaven, the Thirty Years' War led to the establishment of the Protestant creed (as a result of the Westphalian Peace Treaty in 1648, which determined that each nation was sovereign and allowed to decide its own religion). At this time Protestantism was also introduced to North America by the Pilgrims. Under the Thirteenth Heaven the conversion to the Christian faith of all the Americas was completed, and Christianity, through colonialism and missionary activities, spread to other parts of the world as well, notably Africa, Siberia, and Oceania.

CHRISTIANITY IN CONTEXT

Christianity expanded its range of believers and institutional power especially during the odd-numbered Heavens Nine, Eleven, and Thirteen of the Great Cycle, the very Heavens that we have previously found to be dominated by the World Tree, the Invisible Cross in the Heavens (see fig 3.11a, page 46). Not surprisingly, then, in the later Heavens of the Great Cycle, the papacy in Rome, Calvin in Geneva, and Luther in Worms (all residing under the midline) formulated the most influential doctrines of the Christian faith. Because of the creative tension induced by the Invisible Cross, these doctrines first originated along one of its arms and have since spread from there. Thus it is not really the Christian Church that has spread the cross, but the Invisible

Cross that has spread the Christian Church. Or perhaps it would be more accurate to say the Christian Church has been able to surf on the waves generated by the World Tree.

This evolution of the Chaldean-Jewish-Christian tradition clearly expresses at least three tendencies.

First, it is a progression from polytheist to monotheist systems of belief. Whereas in ancient Sumer, at the beginning of the Great Cycle, the Creator God shared his place with many other deities, the roles of the latter have progressively diminished with every odd-numbered Heaven.

Second, the Great Cycle has increasingly developed a belief in a God who is universal—the God of all human beings. When the cycle began, each people had its own gods and particular ways of worshipping them, but gradually religions evolved that sought to include all of humankind, regardless of nationality.

Third, there has been an increasing tendency to worship a transcendent God. At the beginning of the cycle people worshipped many kinds of gods that were seen to animate physical phenomena—city gods, harvest gods, house gods, and so on—but toward its end a view of God as transcendent to all the physical manifestations of Creation has come to dominate.

The differences between the Catholic and Protestant faiths are most striking with regard to this relationship to physicality. Protestantism, which was not firmly established until the Thirteenth Heaven began its rule, has not seen relics of saints, or other physical manifestations of creation, as holy objects of worship to the same extent as has Catholicism. Rather it has emphasized the direct relationship between the individual and a God in the beyond.

To summarize this progression: The Thirteenth Heaven has generated an increasingly monotheist, universalist, and transcendent view of God. In the Chaldean-Jewish-Christian tradition, as with the development of writing, it is the odd-numbered Heavens of the Great Cycle that, pulse by pulse, have created more and more pronounced expressions of the initial seed.

These seven pulses of duality have done more than favor a monotheist belief in a God of the beyond. In a sense, Protestantism is also an expression of the completion of a process that separated humankind from God and made the experience of a living cosmos disappear. Sensitivity to spiritual energies vanished, and on the level of consciousness the Thirteenth Heaven completed the expulsion of human beings from the Garden of Eden. Thus there was a price to be paid for the abstract left-brain knowledge of the One God: the experience of separation from the divine.

In the Judeo-Christian tradition, this pulsewise development of the separation of humans from the living cosmos throughout the Great Cycle resulted in a conflict between human beings and a feared God who was often perceived as punishing. Because human beings in the ruling creation field (see fig. 3.11a, page 46) experienced themselves as partial, with half their consciousness shut out, they could easily be manipulated into thinking there was something wrong with them. From this arose the notion of "original sin," which was formally introduced by Augustine around 420 C.E. as Tezcatlipoca, the lord of darkness, took up his rule.

Early in the cycle human beings made sacrifices to appease God or to heal the split with him. Later, Christian theology developed a new path for the reconciliation of humanity and God by suggesting that God had sacrificed his only son with such a reconciliation in mind. If we consider that, according to legend, the cross of Jesus at Golgotha was made from the Tree of Life in the Garden of Eden (the World Tree), it becomes possible to understand this fundamental idea of Christianity. The immense appeal that Christianity has had may be because God, through this sacrifice, seemed to be saying: "I am sorry about the duality that I have created for you through the World Tree, but if you believe in me and seek to become like my only son, then ultimately there will be a path to an eternal life, beyond this duality." In this interpretation it makes sense that Jesus, as is implied in the New Testament, was meant to be crucified, thus symbolically sharing humanity's plight caused by the World Tree, but also bridging its duality. Through the

Resurrection a path to an eternal life generated by a timeless cosmic consciousness was revealed.

The reason that myths about a Fall are universal among humankind is that people everywhere were affected by the duality that has ruled during the Great Cycle. The experience of being split led to the idea that there was something wrong either with humans or with God, which was explained by the story of a Fall that occurred at an earlier stage of Creation. The religions have thus evolved in response to a need to deal with the polarity created by the World Tree. To varying degrees, all religions have been influenced by the separation this has caused. In Mesoamerica, for instance, the perceived need to appease the gods led to human sacrifice.

Yet we should note that the very need for religion (in Latin, *re + ligare*, "reconnect") emerges only in civilizations in which direct spiritual contact with the divine has already been partially lost. Religions develop rituals to help human beings rekindle the lost contact with their divine source. Such rituals, however, are palliative at best. During periods in which the global field of consciousness has prevented direct contact, religions tend to lose their original meaning.

GOD'S CREATION EQUATION

In the previous chapter we saw that the evolution of human consciousness and history is really a wave movement, and that this movement is generated by the World Tree. We have now also seen that prophets have been inspired to step forth, especially at the beginning of odd-numbered Heavens, with significant new teachings regarding the nature of God and the purpose of creation. These synchronies between the dispensations of prophets and the beginnings of odd-numbered Heavens are particularly evident with Moses, Isaiah, and Jesus/Paul at the beginnings of the Fifth, Seventh, and Ninth Heavens, respectively. Why were the most important steps in the evolution of the Chaldean-Jewish-Christian tradition taken at the beginnings of the odd-numbered Heavens?

Part of the explanation is given by the Mayan creation story, which said that when the World Tree was raised, the light would enter. In all odd-numbered Heavens (see fig. 3.11a, page 46) there is a yang aspect. During the Great Cycle there are thus seven Heavens during which divine light allows prophets and their followers to see divine reality in a new way, and six intermediate periods of darkness (see fig. 3.11b, page 46) when the Invisible Cross is not apparent.

The conclusion comes naturally: The seven odd-numbered Heavens and the six even-numbered Heavens of the Maya are none other than the seven Days and six Nights, respectively, of God's creation. These are described in the Book of Genesis and are also symbolically represented by the seven candlelights of the Jewish Menorah (fig. 4.2).

Figure 4.2. The Jewish menorah (seven-armed candelabra) is a well-known symbol of the work of God's creation completed in seven Days and six Nights.

The wave pattern of the Great Cycle can be formulated as a very simple equation:

$$\text{seven Days} + \text{six Nights} = \text{Thirteen Heavens}$$

This equation is the "solution" to the Mayan calendar, a solution that provides the key to its demystification and allows its mathematical language to be understood by other traditions or lines of human thought than its own. That such a solution exists should not come as a surprise. After all, if inhabitants of the Western and Eastern Hemispheres alike are one and all part of the same creation, and if God's creation is universal, common traits must exist in the ways creation has been understood on the two different continental blocks of the planet. If we are all part of the same creation, children of the same God, then creation stories in different parts of the world must, all in their own ways, seek to describe the common reality. The previous equation is an example of this. It is the crucial bridge between, on the one hand, the Mesoamerican calendar with its Thirteen Heavens, and on the other a few billion people—Jews, Christians, and Muslims alike—whose holy scriptures say that God created the world in seven Days and six Nights.

Incidentally, it is somewhat ironic that the Mayan books burned by the Catholic friars as the "work of the devil" can now be seen to have contained the temporal maps for the development of the divine creation. The Mayan knowledge did not contradict the Jewish-Christian view in the Books of Genesis and Revelation (where the same theme recurs). On the contrary, with their calendar the Maya developed a more exact science of creation that makes it understandable through the events of history. Even more ironically, the Mayan calendar now proves the existence of the same God that the Spanish bishops wanted the natives to believe in.

But then, is there any direct evidence from the Mayan culture that they were aware of this equation? Considering how preoccupied the Maya were with these types of questions, it would be very surprising if they had not known about it or left some sign of their knowledge. And yes, there is one such sign, which, although it may be unique, is all the more telling that the Maya, at least those under the Eleventh Heaven in Chichén Itzá, had identified the Thirteen Heavens with the seven Days and six Nights of creation. This sign is the descent of the Feathered Serpent along the staircases of the Pyramid of Kukulcan, which becomes visible only at the spring and autumn equinoxes (fig. 4.3).

There, in the most spectacular way possible to the ancient Maya, the rhythm of the process of creation is made evident as seven triangles of light on the back of the serpent. Together with the six triangle-shaped shadows that are formed from the terraces, these seven triangles of light give rise to a wavelike pattern consisting of thirteen different triangles, the Thirteen Heavens. In the otherwise "silent" city of Chichén Itzá—

Figure 4.3. At the spring and autumn equinoxes seven triangles of light form along the staircase of the Pyramid of Kukulcan in Chichén Itzá.

which generally lacks written inscriptions—the descent of Quetzalcoatl along the Pyramid of Kukulcan was the most spectacular and educational way to convey, through the architecture of the pyramid, that seven scales of the serpent were light and six were dark. That the Mayan astroarchitects were able to plan the construction of the pyramid in such a way that this message of the wavelike descent of the Heavens to earth was visible only at the equinoxes is truly astounding. It tells us that this pattern of light, which we can still observe today, was considered a most profound revelation.

Through this, the epic of creation becomes complete. The triangles on the side of the Pyramid of Kukulcan bring the creation stories of the Old and New Worlds together. What the descent of the Feathered Serpent reveals is that the Maya, who built the pyramid at Chichén Itzá in the sixth Day of the Great Cycle, were aware of history as a wavelike process created by seven pulses of light alternating with six periods of darkness.

The equation on page 73, as simple as it is, also provides a bridge to empirical science, since the existence of the Thirteen Heavens ruling the thirteen baktuns can be empirically verified. We have already seen a few examples. Thus we have, perhaps for the first time, proof of the existence of God that is not merely based on feelings or abstract philosophical reasoning, but on the Mayan calendar matched to the factual basis of modern historical research and science. Anyone with access to an encyclopedia and the timeline of the Mayan Great Cycle can study how divine creation has played out in reality (see fig. 4.4), and what influence its Heavens have had on human consciousness. The existence of God has been scientifically proved, and we may now study the actual workings of the cosmic plan.

What is unique in the cosmology of the Maya is thus not the idea that we are living in a divine creation, or even in a divine process of creation progressing through alternating periods of light and darkness. At least on a subconscious level, the adherents of all the major monotheist religions seem to have been aware of this too. In fact, at some level, this awareness seems to have been shared in ancient times by all peoples that have considered the number 7, the number of pulses of divine light among the Thirteen Heavens, as holy. The special contribution of the Maya was to determine precisely the durations of these seven Days and six Nights of the divine process of creation.

The realization that we are living in a creation that progresses and develops according to a specific rhythm has very significant consequences for the way we look at life. The main conclusion is that our lives are indeed lived within a context, a divine creation that has a direction and a higher purpose. To the extent that we are able to align our individual lives with the greater purpose of this ongoing creation, they too will be meaningful. Later we discuss how this may be accomplished.

2721 BCE		1932 BCE		1144 BCE		355 BCE		434 CE		1223 CE		2011 CE
Day 1		Day 2		Day 3		Day 4		Day 5		Day 6		Day 7
3115 BCE		2326 BCE		1538 BCE		749 BCE		40 CE		829 CE		1617 CE

Figure 4.4. The Days and Nights of the Great (National) Cycle related to Gregorian years. The years represent significant shifts (Days/Nights) in the spiritual energy of this cycle. Each of the thirteen steps is a baktun (roughly 394 years).

Another conclusion is that the evolution of our consciousness may not have been under our own control to the extent that many seem to believe, or would like to believe, but rather that God's design controls the evolution of consciousness by invisible mechanisms beyond the reach of our manipulations.

THE RELIGIONS OF THE WEST

The religious beliefs of the Western Hemisphere also display marked development during the Days of the Great Cycle. The oldest known culture of Mesoamerica, for instance, that of the Olmecs of the Gulf Coast of Mexico, began to flourish at the beginning of the Fifth Heaven. The tzolkin itself is known from the midpoint of the Seventh Heaven. The tzolkin then emerged as part of a worldwide burst of religious thought (fig. 4.5) called by historians the Axial Age. This oldest evidence of the use of the Sacred Calendar is from about 600 B.C.E. at the Zapotec site of Monte Albán in Oaxaca, Mexico. At the same time that the Jews came to the understanding that there was one God ruling all of creation, Mesoamericans thus started to chart its pattern in time. This geographical division of knowledge is linked to the fact that through holographic resonance it is the left brain that discerns linear time and sequential order.

Later the cult of the Feathered Serpent, called by the Maya Kukulcan and by the Mexicas Quetzalcoatl, seems to have developed in particular during the days of the Heavens Nine and Eleven. The worship of this deity first became prominent in the huge city of Teotihuacán, outside present-day Mexico City, where a special temple was also dedicated to the Feathered Serpent. In fact, Heaven Nine, which incidentally also saw the emergence of Christianity in the Old World, was seen to be ruled by Quetzalcoatl. Following the collapse of Teotihuacán and the classical Mayan culture, the cult of Quetzalcoatl became very prominent during the Eleventh Heaven, when it played the role of an official religion both in Chichén Itzá and in the Toltec city of Tula.

Tzolkin Chart 3

Mayan Day Signs   **Aztec Day Signs**

Mayan	3115 BCE	272	2327	1932	1538	1144	749	355	40 CE	434	829	1223	1617	Aztec
Imix	3095	2701	2307	1913	1518	1124	730	336	60	454	848	1243	1637	Cipactli
Ik	3076	2681	2287	1893	1499	1104	710	316	79	474	868	1262	1656	Ehecatl
Akbal	3056	2662	2267	1873	1479	1085	690	296	99	493	888	1282	1676	Calli
Kan	3036	2642	2248	1853	1459	1065	671	276	119	513	907	1302	1696	Cuetzpallin
Chicchan	3016	2622	2228	1834	1439	1045	651	257	139	533	927	1321	1716	Coatl
Cimi	2997	2603	2208	1814	1420	1025	631	237	158	553	947	1341	1735	Miquiztli
Manik	2977	2583	2189	1794	1400	1006	611	217	178	572	967	1361	1755	Mazatl
Lamat	2957	2563	2169	1775	1380	986	592	198	198	592	986	1381	1775	Tochtli
Muluc	2938	2543	2149	1755	1361	966	572	178	217	612	1006	1400	1794	Atl
Oc	2918	2524	2129	1735	1341	947	552	158	237	631	1026	1420	1814	Itzcuintli
Chuen	2898	2504	2110	1715	1321	927	533	138	257	651	1045	1440	1834	Ozomatli
Eb	2878	2484	2090	1696	1301	907	513	119	277	671	1065	1459	1854	Malinalli
Ben	2859	2465	2070	1676	1282	887	493	99	296	691	1085	1479	1873	Acatl
Ix	2839	2445	2051	1656	1262	868	474	79	316	710	1105	1499	1893	Ocelotl
Men	2819	2425	2031	1637	1242	848	454	60	336	730	1124	1519	1913	Cuauhtli
Cib	2800	2405	2011	1617	1223	828	434	40	355	750	1144	1538	1932	Cozcacuahtli
Caban	2780	2386	1991	1597	1203	809	414	20	375	769	1164	1558	1952	Ollin
Etznab	2760	2366	1972	1577	1183	789	395	AD 1	395	789	1183	1578	1972	Tecpatl
Cauac	2740	2346	1952	1558	1163	769	375	20	415	809	1203	1597	1992	Quiahuitl
Ahau	2721	2327	1932	1538	1144	749	355	40	434	829	1223	1617	2011	Xochitl

Figure 4.5. The basic pattern of seven Days and six Nights is an integral part of the divine energies of the tzolkin. This pattern can be applied to several different creation cycles, such as the thirteen baktuns (260 katuns, expressed as Gregorian years) of the Great Cycle shown above.

It is not surprising that the cult of Quetzalcoatl developed during odd-numbered Heavens, since this deity was a symbol of the light aspect of the yin/yang duality of these Heavens. More directly, the Feathered Serpent was a symbol of the duality maintaining the distinction between light and darkness; in a sense, it was this duality that created the light. Most World Tree myths, including the Norse and the Jewish, describe a serpent moving up and down the World Tree, and in many cultures the serpent has been a symbol of duality.

I suspect that to the Maya the Feathered Serpent was in fact a similar symbol of an energy spiraling around the World Tree trunk (longitude 12° East) during the odd-numbered Heavens when light fell on the Western Hemisphere. (Hence the Mesoamerican prophecy that Quetzalcoatl would return from the East.) Since the cult of Quetzalcoatl took its most important steps at the odd-numbered Heavens and seems to have appeared clearly for the first time around the beginning of the Ninth Heaven, it has often been suggested that Christ and Quetzalcoatl are the same, divine symbols—or incarnations—of the principle of light emerging at those times. From the perspective developed here, this makes perfect sense.

The Aztecs, in contrast, who wandered into and conquered Mexico from the north during the Night of Heaven Twelve, saw themselves as living in a time ruled by darkness. In their era it was no longer possible to see the divine reality with the clarity that the Maya had previously enjoyed. Nonetheless, as discussed in chapter 6, the Aztecs also had an awareness, expressed through their calendar, of the basic pattern of seven Days and six Nights of creation. Given what we know about the even-numbered Heavens as periods when human creativity flourished and prophets emerged, the characters of the different gods of the Aztecs ruling these Heavens (see fig. 2.2, page 19) also become understandable.

In the West, in the Americas, creation most often seems to have been regarded as the result of teamwork among the gods, who were seen as deities with often humanlike qualities. Originating as it did in the Western Hemisphere, the Mayan creation story was also typically more mathematical and analytical than the creation stories developed by their contemporaries in other parts of the world. Yet the Mayan view was not an analytical viewpoint that detached humans from creation and nature. It was a science of time that saw humankind as an integral part of the web of nature and promised to restore its role as cocreator in the cosmic processes.

THE RELIGIONS OF THE EAST

In the Eastern Hemisphere the midpoint of the Great Cycle around 550 B.C.E. was a point marked by innovative religious thought (fig. 4.1, page 66). At about this time the philosophies of Buddhism, Confucianism, Zoroastrianism, Jainism, and Taoism and the Hindu concept of reincarnation were all conceived. In fact, the dates of birth of both Confucius and Buddha are given as exactly the midpoint of the Great Cycle, 551 and 552 B.C.E., respectively. Even if these dates are mythical, they still indicate at least a subconscious awareness on the part of their followers that this was an important point in time (fig. 4.6).

Generally speaking, Buddhism and some other major Eastern religions also seem to have undergone significant development during the Days of the Great Cycle. In the role played by a single Creator God there is, however, a very noticeable difference between the traditional religions of the East and those of the midline. The role of a Supreme Creator God is clearly more strongly emphasized around the midline as compared to the East, where the belief in a single God was never widespread, at least not in the sense of a personal God. Even if Brahma is seen as the Creator, he is certainly not the only or even the most important god in the Hindu pantheon; and the Unmanifest Supreme Brahman, which does not create, is too abstract to be personal. In the Buddhist tradition a Creator God is not acknowledged at all. Farther east, in ancient China, the cosmos was, along a similar line of thought, believed to have been created by itself rather than by a Creator God.

Islam seems to play an intermediary role between the Great Cycle consciousness of the Center and of the East. It is a profoundly

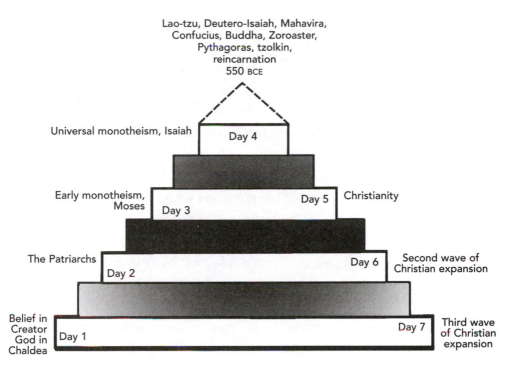

Figure 4.6. The pyramidal evolution of the Chaldean-Jewish-Christian tradition during the Days of the Great Cycle. Notice the worldwide burst of religious teachings at the apex of the seven-story pyramid, corresponding to the midpoint (Day 4) of the Great Cycle (about 550 B.C.E.). The different Nights in the cycle have somewhat different qualities: Nights 2 and 5 (indicated by darker colors) are more destructive or reactionary than the others, which might be described as periods of historic rest.

monotheist religion. Yet its symbols are the crescent moon and the evening star, symbols of the night. Indeed, this religion was founded as the Qur'an was completed in 632 C.E., the exact midpoint of the Night of the Tenth Heaven. Thus its role may be seen as that of bringing light into the darkness. Because it was formulated and shaped by a Night, Islam has primarily spread during the Nights when the Invisible Cross was not as evident.

The message of Islam is somewhat different from that of Christianity, both because it was formulated east of the midline and because it developed during a Night. Islam has reflected an Eastern mentality in its emphasis on collective expressions of worship, such as prayers and pilgrimages, whereas the more Western Christianity has emphasized individual expressions. Since Christianity was conceived during a Day, which is generally a time when human creativity flourishes, its emphasis has been more on the individual crafting his or her own destiny; while that of Islam, conceived during a Night, has been on submission to the will of God.

If we recognize that during the Great Cycle human consciousness has developed as a resonance phenomenon with the vibrations of the World Tree, the historical evolution of the religions becomes understandable. The consciousness of people in Europe and the Mediterranean was much more directly affected by the baktun-based pulsations of the cross than that of populations farther east. In those of Europe and the Mediterranean a belief developed in a personal Creator God as the source of the pulses generated by the World Tree. In the East, in contrast, among Hindus, Buddhists, Confucians, and Taoists, the cosmos was not perceived in this way. This was partly because during the Great Cycle the adherents of these religions and their prophets were living on the dark side of the planet, and partly because they were living at such a long distance from the vertical arm of the cross. It seems that the farther east we go in ancient times, the less belief there was in a Creator God. In Chinese philosophy, for instance, yin and yang seem to have been regarded more as creative forces of nature than as aspects of the divine. By the same token, this also meant that, especially in India, a spiritual tradition with a view of a living, divine cosmos, rather than an abstract and remote Father God, survived. Certainly, if we consider the Great Cycle as part of a cosmic time plan for the enlightenment of humanity, its midpoint deserves to be studied. At the very midpoint of the Great Cycle, Buddha, and his less famous contemporary Mahavira, were considered to have attained the enlightened state. Even if the enlightenment of Buddha was not of the same kind as what people may be approaching today, its link to the temporal development of the Great Cycle is nonetheless very noteworthy. It created a tradition that would have a great influence on thinking, initially in the East and more recently in the West.

In the study of religions we have found that the Mayan calendar is a global calendar that describes the ups and downs of the historical process and the evolution of its religions. The different religions and spiritual traditions are thus results of one and the same divine creation process, and so it is only natural that they all seek to relate to the same spiritual reality. A great power of the Mayan calendar is to be found in its potential for unifying different religious perspectives by bringing about the realization that they all describe this one reality. This allows us to see things from a higher perspective where differences between religions may be transcended.

The higher perspective generated by the Mayan calendar, however, cannot be turned into a new dogma, or a new organized and ritualized religion that seeks proselytes. Rather the Mayan calendar provides a possible framework for the common exploration of spiritual reality by individuals who share a respect for the contributions and views of others. The Mayan calendar, properly understood, is in its essence alien to all fundamentalism, to everyone believing that there is but one true religion that holds the whole truth. Sadly, this is why the Catholic priests burned all the books of the Maya: They threatened the monopoly on the truth that Catholicism at the time was striving to establish. Today, however, increasing numbers of people are turning away from organized religion out of a desire to seek the truth for themselves. This facilitates the revival of the Mayan calendar on a worldwide scale among all those aspiring to the unity of humanity.

(1 + 12) IS NOT THE SAME AS 13

It may now be clear that the reason that certain numbers play such a great role in Mesoamerican cosmology is that these numbers are very important in the greater creation scheme. There is something to be learned about how different religions look upon creation from the numbers they consider holy. Let us now consider the difference between the numbers 12 and 13, and the significance these have been given in different cultures. In Mexico among the Maya and the Mexica, the number 13 was the most consistently upheld as a holy number. In Mesoamerica it formed part of the thirteen-day count and the Thirteen Heavens of creation. Moreover, kings would typically celebrate their thirteenth tuniversaries as important occasions in their lives.

In the rest of the world, in contrast, the importance of the number 13 has almost been hidden, and in recent centuries it has even been considered unlucky (Friday the thirteenth). Even earlier the number 12 often overshadowed the number 13, and in the European tradition we have to add one (Jesus) to the twelve disciples to arrive at the number 13 in the Babylonian-Greek system of astrology. One (ourselves) is also added to the twelve zodiacal signs to arrive at thirteen. In the Chinese astrological cycle of twelve years, it is even questionable whether there is an individual in the center to add. It thus gives even more emphasis to the number 12.

So while the number 13 was revered in Mesoamerica, the rest of the world placed emphasis on 1 + 12. This concealment of the number 13 in the Old World is also evident when we examine the description of the process of creation in the Book of Genesis. There it is said that on the seventh Day God rested, implying a distinction between the twelve periods of active divine creation and the thirteenth.

Although the difference between 13 and 1 + 12 may seem insignificant on first glance, it highlights vast differences between philosophical views of what it means to be a human being and the human role in creation. If we use 1 + 12 as the mathematical organizing principle of our religious-cosmological worldview, it means that the individual (one) is regarded as separate from the rest of creation (twelve). In such a view, the individual is also regarded as the center of creation, as a chosen one. It is a central tenet of the Christian religion that its adherents should aspire to emulate the ideal of Jesus Christ in their thoughts and deeds. This would then mean becoming a chosen one, who stands out from the other aspects of creation (twelve).

Of course, most people would say that 1 + 12 equals 13, but in the

Mesoamerican view numbers are not really quantities but qualities. Thus if one is seen as different from the other twelve, the numbers cannot really be added. Although 1 + 12 mathematically equals 13, the supreme number of creation, the 1 + 12 cosmology has led us astray. The 1 + 12–based astrological systems of the Center and the East, for instance, have never been proved to correspond to a tangible reality or to clearly evident manifestations of the evolution of consciousness. In contrast, the 13-based "astrology" of the Maya is the only such system that has been proved to have a basis in reality in its correspondence to the changing energies of human history. Yet, paradoxically, it is probably the least known of all divinatory systems. (See fig. 4.7.)

Since approximately the tenth century C.E., the Catholic Church had assimilated Babylonian astrology (which in the modern world has become known simply as "astrology"). It is thus not very surprising that this type of physically based astrology was forced upon the Maya together with Christianity, since the two are based on the same 1 + 12 philosophy. Related to this is the idea prevalent in recent decades of the earth going through different ages—Aquarius, Pisces, and so on—because of its precessional movement. This in turn is based on a geocentric view of the earth as a specially chosen planet (one) set apart from the surrounding galaxy (twelve).

Despite its message of love and compassion, Christianity is also a midline religion with a 1 + 12 philosophy, which in and of itself has

West	Center	East
Analysis		Synthesis
Individuality		Collectivity
Plurality of Creator Gods	Single god	No Creator God
13	1 + 12	1? + 12
Thirteen heavens	Individual + twelve zodiacal signs; Jesus + twelve disciples	Twelve years

Figure 4.7. Traits of the cosmologies of the West, Center, and East

tended to set the individual off from the universal web of creation of which he/she is a part. This is one aspect of the religious intolerance that has often been typical of Europeans (midliners), whose dominating idea has traditionally been that there is only one right way, which is to be imposed on others (twelve). If instead we base our cosmological system on thirteen numbers, as did the Maya, then the human being may be regarded as an integral part of the web of creation. Cosmological systems based on 1 + 12 mean that human beings are seen as separate from creation and from nature, while those based on 13 mean that they are recognized as part of it. It may well be because creation philosophies based on the number 13 do not offer humans a special position that the number 13 has come to be regarded as bad luck in European culture. (This negativity was later exported to the United States, where hotels do not always have a thirteenth floor or a room number 13).

IN LAK'ECH: I AM ANOTHER YOU

The conflicts between the European Catholics and the indigenous populations of the Americas—the original Westerners—were not only conflicts of race but of mentalities and calendars. The Europeans had been driven from the planetary midline by the yin/yang polarities that were introduced there. This gave rise to an expansive creative tension that ultimately was the reason Europe became the global center of colonialism. On the other side of the coin, Europe brought the world together. Yet the conquistadors were living in a 1 + 12 reality, which led them to break the web of creation and subjugate other peoples to their own domination. One plus twelve is symbolic of separation from the true source of creativity, and it strengthens a view of the universe and its living parts as something separate from oneself, something that is to be conquered and controlled.

We can now begin to appreciate how important the calendrical system developed in Mesoamerica may be for the future of humanity. Calendrical and astrological systems based on the number 12 may lead us to deny the complex web of creation and nature of which we are all part. The use of a calendrical system based on the number 13, in contrast, leads to resonance with nature and the cosmic processes. There is a great difference in outlook on the world between a system in which one is considered as apart and separate and a system where all are seen as equals. The return of the Sacred Calendar thus means the strengthening of a process where people are seen as equals in the web of creation.

Love means relationship, web, participation, equality, and a sense of community. Love is thirteen. The Maya have long used a special expression for greeting others: *In Lak'ech,* which means "I am another you." A whole philosophy has been developed around these beautiful words of recognition, which are applied not only to human beings but also to animals, flowers, stones, and spirits. It means that we are not separate; we are all part of the same web; and if any part of this web is hurt, the rest will suffer. The use of the tzolkin and its thirteen-day count is part of this In Lak'ech philosophy. The use of the Sacred Calendar amounts to an act of reverence for the web of nature and creation.

5

The Nine Underworlds

EXPANDING LEVELS OF CONSCIOUSNESS

We have now looked at the Great Cycle in some detail and seen how the Thirteen Heavens generated by the World Tree have inspired human creativity as well as our understanding of the divine. In Mesoamerican mythology, however, there were not only Thirteen Heavens, but also Nine Underworlds, and we shall now explore the latter.

Although it is not possible to gain a detailed understanding of the origins of these Underworlds, we may surmise that they are nine sequentially activated frames of consciousness mediated by the earth's inner core.

Today we know that the world did not begin only five thousand years ago with the emergence of the first written language, the first pyramids, and the first nations centered on pharaohs. Modern science has shown beyond a doubt that the present universe came into existence much earlier than when the Great Cycle started. It was born some 15 billion years ago as matter was first created from light in the so-called Big Bang. Since then, our galaxy, solar system, and planet with its biological organisms have all come into existence. Hence the Great Cycle did not begin from nothing. It had considerable previous evolution to stand on.

How does the Mayan calendar fit with this great age of the universe? Interestingly, the Maya were aware that the world was much older than 5,125 years. On a stele discovered at the ancient site of Coba on the Yucatán Peninsula (fig. 5.1), the creation date of the Long Count is placed in the context of a hierarchy of creation cycles of 13×20^n (13 times 20 multiplied n times by itself) tuns. Thus there is a whole series of creation cycles like the Great Cycle (which itself is 13×20^2 tuns). Creation may then be regarded as a composite of several creations, each being built on top of another to form a pyramidal structure (fig. 5.2), and of these cycles the Great Cycle is but one. Today we can still see this

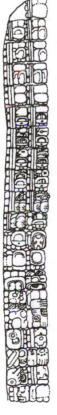

13 alautuns

13 kalabtuns

13 baktuns

0 tun

0 kin

13 hablatuns

13 kinchiltuns

13 pictuns

0 katun

0 uinal

Figure 5.1. Stele from Coba with the beginning dates of several creations (Underworlds). Although the stele includes even longer periods (up to a million billion times the known age of the universe), the periods for which the Maya had names cover the known evolution of the universe (16.4 billion years).

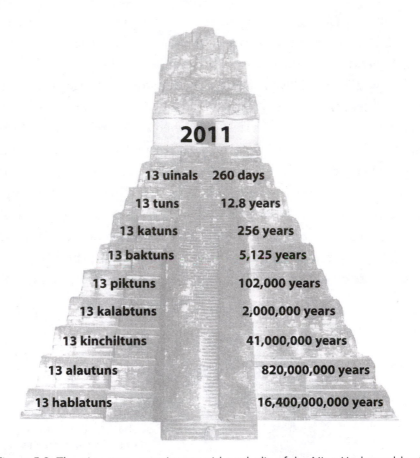

2011	
13 uinals	260 days
13 tuns	12.8 years
13 katuns	256 years
13 baktuns	5,125 years
13 piktuns	102,000 years
13 kalabtuns	2,000,000 years
13 kinchiltuns	41,000,000 years
13 alautuns	820,000,000 years
13 hablatuns	16,400,000,000 years

Figure 5.2. The nine-story cosmic pyramid symbolic of the Nine Underworlds of creation. Each of these develop a certain level of consciousness, and all reach their completion on the day 13 Ahau, October 28, 2011.

idea of several different creations in ceremonies of the Maya, where, for instance, monkeys from a "previous creation" are sometimes enacted. Each of the Nine Underworlds of Mesoamerican mythology is a different "creation" generated by a cycle twenty times shorter than the one it was built on. This is why the most important of the Mayan pyramids—the Temple of the Inscriptions in Palenque, the Pyramid of the Jaguar in Tikal, and the Pyramid of Kukulcan in Chichén Itzá–were all built as hierarchical structures with nine levels.

Although these Underworlds (fig. 5.2) have sometimes been translated as "hells," I am convinced that this is misleading. It does not make sense that the Maya had built their nine-story pyramids in honor of some "hells." Underworlds are related to the sequentially activated crystalline structures in the earth's inner core. (The origin of the Christian hell was the Norse Earth Mother Hel, who ruled the Underworld and was portrayed as a fearsome force in the patriarchal Great Cycle.) The beginning dates of the creation cycles generating these nine Underworlds, and some concomitant events (seeds)—which, in accordance with modern scientific dating, occur at the beginnings of the Underworlds—are summarized in figure 5.3.

The Great Cycle, creating the Sixth Underworld and generating a national frame of consciousness, to which we have until now given all

Underworld	Spiritual Cosmic Time	Physical Earth Time	Initiating Phenomena	Scientific Dating of Initiating Phenomenon
Universal	13×20 kins	260 days	?	
Galactic	13×20^0 tuns	4,680 days (12.8 years)	?	
Planetary	13×20^1 tuns	256 years	Industrialism	1769 CE
National	13×20^2 tuns	5,125 years	Written language	3100 BCE
Regional	13×20^3 tuns	102,000 years	Spoken language	100,000 BCE
Tribal	13×20^4 tuns	2 million years	First humans	2 million years
Familial	13×20^5 tuns	41 million years	First primates	40 million years
Mammalian	13×20^6 tuns	820 million years	First animals	850 million years
Cellular	13×20^7 tuns	16.4 billion years	Matter; "Big Bang"	15–16 billion years

Figure 5.3. The durations of the Nine Underworlds in both spiritual and physical time and some of their initiating phenomena

our attention, is thus just one among several such creations, or Underworlds. This particular Underworld was built on the foundation of five lower Underworlds that had generated more limited frames of consciousness. The nine-story Mayan pyramids are thus telling us that consciousness is created in a hierarchical way and that each Underworld stands on the foundation of another.

Before continuing we should summarize the various time periods that are part of the tun-based (360-day) calendrical system of the Maya (fig. 5.4). Each of the nine major creation cycles is based on a period that is a multiple of the tun. The tun is multiplied by twenty, different numbers of times, in such a way that a hierarchical system of time periods is generated. (Note the exception, however, that there are eighteen, rather than twenty, uinals in a tun).

Mayan Name of Period	Spiritual Cosmic Time	Physical Earth Time
Kin	1 kin	1 day
Uinal	20 kins	20 days
Tun	1 tun	360 days
Katun	20 tuns	7,200 days or 19.7 years
Baktun	20^2 tuns	144,000 days or 394 years
Piktun	20^3 tuns	2,880,000 days or 7,900 years
Kalabtun	20^4 tuns	158,000 years
Kinchiltun	20^5 tuns	3.15 million years
Alautun	20^6 tuns	63.1 million years
Hablatun	20^7 tuns	1.26 billion years

Figure 5.4. Summary of time cycles linked to the tuns (360-day years) and their corresponding durations in physical time

Each of the Nine Underworlds is developed through a sequence of Thirteen Heavens with seven Days and six Nights, beginning with thirteen *hablatuns* in the Cellular Underworld and continuing upward through *alautun, kinchiltun,* and so on. Hence at every new level the Thirteen Heavens (of a twenty-times-shorter duration than the level below) serve to develop a higher frame of consciousness. Thus each Underworld is associated with a certain frame of consciousness: Cellular, Mammalian, and so forth. Each Underworld is also associated with a certain frequency of creation, the frequency with which the energy changes take place. At all levels, the energy changes are the results of alternations between seven Days and six Nights. As an example, the sixth level (the Great Cycle, or National Underworld), ruled by a commonly mentioned deity by the name of Six-Sky-Lord, has a baktun period (394 years) of energy changes that serves to develop its national frame of consciousness. In figure 5.5, the different energy changes of the Regional, National, and Planetary Underworlds, each taking place with a certain frequency, are shown in a correct comparative scale. There we see how the National Underworld began a given time after the seventh Day of the Regional Underworld was established, and then in turn how the entire Planetary Underworld falls within the seventh Day of the National Underworld. In the scale used in figure 5.5 the longest creation cycle, the Cellular Underworld, would stretch some twenty-five miles (forty kilometers). This is a good illustration of the phenomenon that evolution is accelerating.

As of this writing (2003), we have come to the Third Day of the creation of the Eighth Underworld, the Galactic (see fig. 7.1, page 138), which in turn stands on the Planetary Underworld. There are now only two more Underworlds to go—the Galactic (12.8 years) and the Universal (260 days)—until we reach the highest level of creation. In this hierarchical structure, the different frames of consciousness (Regional, National, Planetary, and so forth.) do not replace each other, nor do they follow one another. Instead they add to each other, so that the creation of all the Underworlds, and thus the climb of the nine-story cosmic pyramid, will be completed at the same time: October 28, 2011,

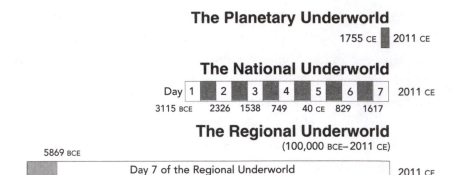

Figure 5.5. Relationships among the Regional, National, and Planetary Underworlds presented in an accurate time scale and illustrating the increase of the frequency of change. All cycles contain seven Days and six Nights and carry the development from seed to fruition. On this scale the lowest Underworld (the Cellular) would measure some twenty-five miles.

in the Gregorian calendar. As we will shortly see, each Underworld is also associated with a specific yin/yang polarity.

Creation thus brings about the evolution of the cosmos through nine distinct Underworlds, going from the lower Underworlds, where the evolution of consciousness is manifested in physical ways—such as galactic matter and biological species—to the increasingly more ethereal or spiritual expressions of the higher Underworlds. Looking at the ascent of the cosmic pyramid from the perspective of modern knowledge, we may identify a sequence of different frames of consciousness, each associated with a particular Underworld. Because these frames expand from the Cellular to the Universal, the purpose of creation seems with every Underworld to be to expand and elevate this frame, ultimately making possible the advent of the Universal Human.

THE BIG BANG

The first of the Nine Underworlds, the one generated by the thirteen hablatuns, started 16.4 billion years ago. This is very close to the point in time estimated for the Big Bang, the term modern physicists use to describe the very beginning of creation. This First Underworld provided the material basis for the rest of creation: matter, galaxies, solar systems, and cells. The reason this Underworld has been called Cellular in figure 5.3 (page 191) is that the higher cells, emerging at the beginning of its seventh Day, represent its highest level of consciousness. The other Underworlds have been given names in an analogous manner.

That the development of the First Underworld begins with the Big Bang has very far-reaching consequences that should not pass unnoticed. First, here we have definite proof that the tun-based calendrical system of the Maya is not based on astronomical movements or biological cycles. This calendar system can be used to describe the process of creation from a time when no solar systems or even galaxies existed. The tun-based calendrical system thus describes divine creation processes that are primary to any of their material manifestations and is conceived on a very grand scale. The processes that the tun-based system describe go back to a time when God's Word began this creation or, in Mayan terminology, when the First Father raised the vibrating Universal World Tree. (The World Tree discussed in chapter 3 is only the local representative on our own planet of the Universal World Tree.)

Of course, it is not likely that the Maya had the same type of knowledge about early creation as today's physicists, who are able to describe events second by second through quantum mechanical considerations. In any case, such physical knowledge may not have been of much interest to Mayan sages, because some of it only serves to obscure the fact that we are living in a creation with a divine purpose. The Maya did know, however—and to this they were guided by their deep understanding of the Sacred Calendar—that there is a special rhythm with which creation unfolds, and that there were nine different major creation cycles, each creating a specific Underworld of consciousness.

Second, since the First Underworld covers all time since the Big Bang, everything that we now know to exist—all of creation—is a result of the creation processes generated by these Nine Underworlds, each developed through Thirteen Heavens. As is shown in figure 5.6, the types of phenomena that each of these Underworlds develop are very different and specific to each Underworld.

Underworld	Duration	Level of Consciousness Phenomena Evolved Frame of Life
Universal (Ninth)	13 uinals	**Evolution of cosmic consciousness** No limiting thoughts, timelessness No organizing boundaries
Galactic (Eighth)	13 tuns	**Evolution of galactic consciousness** Transcending material framework of life, telepathy, living on light, genetic technology Organized in galaxies
Planetary (Seventh)	13 katuns	**Evolution of global consciousness** Materialism, industrialism, Americanism, democracy, republics, electrotechnology Organized in planets
National (Sixth)	13 baktuns	**Evolution of civilized consciousness** Written language, major construction, historical religions, science, fine art Organized in nations
Regional (Fifth)	13 pictuns	**Evolution of human consciousness** *Homo sapiens* with ability to make complex tools, spoken language, art, early religion Organized in regional cultures
Tribal (Fourth)	13 kalabtuns	**Evolution of hominid consciousness** Human beings (*Homo*) who make complex tools and have rudimentary oral communication Organized in tribes
Familial (Third)	13 kinchiltuns	**Evolution of anthropoid consciousness** Lemurs, monkeys, Australopithecans with the ability to walk upright and use tools Organized in families
Mammalian (Second)	13 alautuns	**Evolution of mammalian consciousness** Evolution of multicellular organisms, sexual polarity, a continental structure and plant kingdom to support higher life Higher mammals
Cellular (First)	13 hablatuns	**Evolution of cellular consciousness** Step-by-step evolution of the physical universe: galaxies, stars, and planets; evolution of chemical elements Higher cells

Figure 5.6. The durations and chief phenomena developed by each of the Nine Underworlds. We have reached the seventh Day in the seven lower Underworlds and will enter the Ninth and final Underworld in February 2011.

THE HUMAN BEING

Although it cannot be presented in detail here, it can be proved by studying the four lowest Underworlds that living organisms have not come into existence through mere chance, as the neo-Darwinist school currently ruling biology would have it. Instead biological evolution is a result of the divine creation cycles described by the Mayan calendar. Built on the biological evolution that was brought about by three lower Underworlds, the fourth of these, the Tribal, began with the emergence of the first human beings. According to fairly unanimous anthropological research, humans first appeared in central Africa about 2 million years ago. Somewhat east of the center of the World Tree the oldest remains of true members of the human species, *Homo habilis,* have been found, and their hominid predecessors have been found even closer to the midline (in Chad). We may then understand the ancient Mayan myth according to which the first human being was born out of the World Tree. (The same idea, incidentally, recurs in Nordic myth, where the first human being, Ask—"ash"—was born from the World Tree, the huge ash Yggdrasil.)

In the Underworlds built on top of the Fourth Underworld, human beings have come to creatively transform the matter already created in the lower Underworlds; and then, increasingly with every higher level

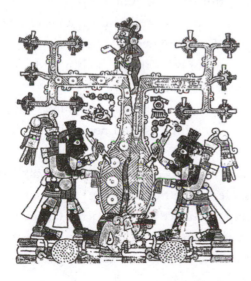

Figure 5.7. A human being reborn from the World Tree

of consciousness, they have developed more fully into cocreators with the Creator. In this regard it is the Fifth (or Regional) Underworld, which started about 102,000 years ago, that represents the most important turning point. In this Underworld human beings began to speak a diversified language, and to create sophisticated tools and art.

Obviously, the birth of the universe, the Big Bang, and the birth of human beings are major events in cosmic evolution. If we then also consider that the emergence of the first human civilizations coincides with the beginning of a major creation cycle, the Great Cycle, it becomes evident that each of the Nine Underworlds indeed serves to develop a very significant new level of consciousness (fig. 5.6, page 96).

From such examples we realize that the Mayan calendar is truly a means for building bridges of understanding—not only, as seen in the previous chapter, between the different religions and spiritual traditions of the world, but also between spirituality and science. This is because the validity of the divine creation cycles may be verified by scientific facts about evolution. No longer is there a contradiction between evolution and creation. These are only two different ways of looking at the same process.

The human being is thus himself or herself a result of cosmic creation that in many different ways has been patterned by the tzolkin. Thus, for instance, there are thirteen different major joints in the human body, seven above and six below the waistline, and we have a total of twenty toes and fingers. There are twenty different amino acids that build our proteins, the agents of cellular metabolism, and 260 different distinct cell types in the human being.

Some of these numbers are specific to humans. No animal has thirteen estrus cycles in a year or 260 cell types, and not all have thirteen joints and a sum of twenty toes and fingers. This very likely means that the human being, more than any other species, embodies the tzolkin pattern of creation in its design. Hence the human being may to an exceptional degree be able to develop resonance with the Sacred Calendar and *tune* in to the cosmic creation cycles that follow its basic energy pattern. This resonance was behind the ancient Mesoamerican thought that there

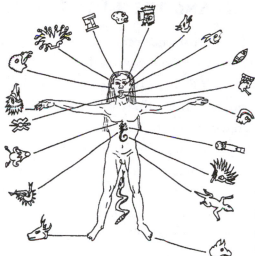

Figure 5.8. Relationships among the various body parts and the twenty day signs.

are links between the energies of the day signs and the body parts of the human being, a concept essential to their medicine (fig. 5.8).

We should also consider another very important parallel between the human being and the Mayan calendar system. The phases humans go through in their development as individuals reflect the creation of the nine levels of consciousness of the cosmos. The parental orgasm and subsequent conception—the unification of the invisible male and female forces of the cosmos—is parallel to the Big Bang, which led to the emergence of the first cells through the working of the Cellular Underworld. As the human fetus goes through the various stages reminiscent of fishes and animals, its development mirrors that of what was created during the Mammalian Underworld, and so forth; until at birth and in early childhood human consciousness corresponds to the enlightened consciousness of the Stone Age people of the Regional Underworld. From ages three through seven years, as a child learns to read and write, starts seeing differences between good and evil, and starts to apply judgments of himself or herself and others, he or she will, as a natural development, experience a Fall. This corresponds to attaining the dualist consciousness of the National Underworld, and so on. We will return to our individual climbs through the Underworlds in chapter 8.

THE THOUGHTS OF GOD AND THE HUMAN AGENT

Einstein once said that his interest in science came out of a desire to understand the thoughts of God. How then may the Mayan calendar help us describe the thoughts of God? Figure 5.9 is a model for the holographic transmission of information from the cosmos to ourselves. Because a spiral galaxy is separated into two different hemispheres, as are our brains, it is possible for us, through holographic resonance, to receive the thoughts of God, the cosmic information generated by a Universal World Tree, which use both the galaxy and the earth as relays. Part of the explanation for why the universe is dominated by polarities—polarity of genders, polarity of brains, and polarity of consciousness—is that these polarities meet the need for a mechanism to transmit information from the divine source. Resonance presupposes polarity, and because of the polarity existing between the hemispheres of our brains, they can serve as antennas for the reception of new information.

It is also entirely possible that the sun serves as a relay for creative cosmic information to the earth. Because of the speed of the sun's rotation around its own axis, a spot on the sun faces the earth for approximately thirteen days. The trecena may thus be a reflection of an invisible World Tree on the sun, which results in the earth being exposed to changing energies over a thirteen-day period. These may be mediated partly by the interaction of the solar wind with the various concentric layers of the earth (chapter 3). It may not be a coincidence that the solar surface temperature is the same as the temperature at the earth's core (6,000°).

In the view of our existence presented here, the human brain serves as an interface for the transmission of information from the nonphysical Heavens to the physical domain. The brains of human beings are "channeling" not only the information, but also the means for processing it. The periods when the "downloading" of such programs are favorable are described exactly by the Mayan calendar. This is most easily demonstrated by the fact that there are a number of aspects of hemispheric preference that operate in parallel on the global and human scales. Examples are language and writing skills, which in more than 90 percent of people are developed by the left hemisphere of the brain. This is parallel to the capacity of humankind as a whole to communicate in writing, which developed as a result of a series of global spiritual pulses (Days in the Great Cycle) where the light shone on the Western Hemisphere of the planet. (Combine fig 2.7, page 28, with fig. 3.11a, page 46, and fig. 5.9!)

Mathematics (which is not discussed in this book) follows the same scheme of evolution, since it is an abstract way of thinking mediated by

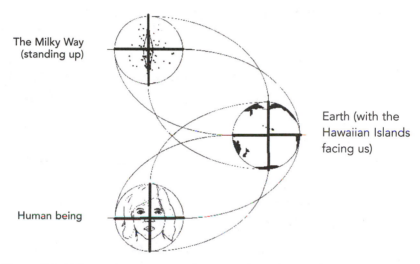

The Milky Way
(standing up)

Earth (with the
Hawaiian Islands
facing us)

Human being

Figure 5.9. Holographic resonance between the galaxy and a human being,
with the earth serving as a relay

the left hemisphere of the brain. It is little wonder, then, that during the past four hundred years (the Thirteenth Heaven) capitalism—the economic system based on the abstraction of values to quantities—has had the West, initially the Netherlands and the United Kingdom and later the United States, as its driving force. It is also little wonder that the ups and down of this abstract economy very strictly reflect the energy changes of the Mayan calendar. Hence economic cycles are also reflections of the cycles of consciousness emanating from the global World Tree. (Appendix A has further information on economics.)

Writing and abstract calculations are only two examples of the hemispheric preference of information relayed by our brains from the nonphysical domain. There are numerous others. The thoughts of God provide the spiritual archetypes that are mediated by our brains according to the exact time schedule of the Thirteen Heavens and Nine Underworlds. These may be visualized as cosmic projections by the light of God.

As presented here, the Mayan calendar is fundamentally a time schedule for the evolution of consciousness. In *The Mayan Factor*, José Argüelles pioneered this basic idea. Argüelles was the first in modern times to take the symbolism of the tzolkin seriously, suggesting that the Day Lords exerted archetypal influences of a spiritual nature on the course of human history. Until this writing, most later researchers who have studied the Mayan calendar have focused instead on the supposed end date of the Great Cycle, which has produced a much too limited view of this calendar system. Inferences about end-date events of a physical nature have, I believe, tended to belittle the enormous explanatory power that the Mayan calendar has proved to have as a description of the evolution of consciousness. This has meant that the whole point of the path toward enlightenment has been lost.

Especially in the West, despite a general movement in the direction of a holistic worldview, many metaphysically oriented people still rely on such physical explanations for what is happening in the world. Rather than recognizing consciousness as primary to matter and that the evolution of consciousness drives all things, many still seek to retain some last straw of a physical foundation, whether it is the photon belt, the precession of the earth (nowhere in the Mayan sources is a 26,000-year cycle mentioned), hypothetical planets such as Niburu or Maldek, a magnetic pole shift, or something similar to explain the course of events. With a recognition that this ongoing creation goes back 16 billion years, prior to the existence of anything material, we would have to recognize that a resting consciousness, an all-pervading intelligence, is ultimately the source of all evolution.

The physical phenomena proposed in the New Age community are, of course, always different from those accepted by the academic community (otherwise they would lose their mystical appeal). The former often share with the latter, however, this same inverted cause-effect scheme that matter is primary to consciousness. This scheme requires that an acceptable theory of science ultimately be explicable in terms of matter. In the current Galactic Underworld, however, more favored is a perception of the world in which all things in existence are seen as products of an ongoing evolution of consciousness. This evolution is generated by the invisible World Tree, which, as it is said in Mayan myth, existed before everything else. For those willing to make the jump to this worldview, the world will be much easier to understand.

Of course, this is exactly the same basic premise that was pointed out earlier regarding the Mayan calendar system itself. To understand the cosmic plan, we need to make the distinction between spiritual and physical time cycles; all attempts to subordinate the evolution of consciousness to astronomical cycles fail to provide the necessary understanding. That the Mayan creation cycles go back to a time when there was only a resting consciousness should make it immediately clear that this resting consciousness is primary to all material manifestations of creation.

Transmission of cosmic information through the ether is apparently not limited by the speed of light, and we must thus assume that the changes in yin/yang polarity occur in synchrony in all the structures of the universe: galaxies, solar systems, planets, organisms, cells, atoms, and nucleons. The information shaping consciousness is disseminated through holographic resonance, which also may be called fractal expansion. Ultimately, all consciousness shifts are generated by the Universal World Tree, which presumably organizes the galaxies of the universe into one unified structure that is not yet clear to us. (This is because the constant speed of light limits our ability to see the whole universe. Unfortunately, scientists often base their discussions of an expanding universe on an assumption that the part we see is the whole universe.) At the center of the Universal World Tree there may be what in esoteric traditions is called the Central Sun. For the spiritual seeker, an understanding of the nature of the ethereal yin/yang fields in physical terms may not be all that important. It is important, however, to learn enough about the Mayan calendar to recognize that this describes the evolution of all things and to understand its prophetic implications.

This Universal World Tree is, then, identical with what in the Mayan tradition is called Hunab-Ku, sometimes referred to as the One Giver of Measure and Movement, but certainly better translated as the One Source of Limits and Energy (fig. 5.10). The limits are actually the boundaries creating yin/yang polarities (on our particular planet, for example, the planetary midline). The energy is the creative energy emanating from the Central Sun. As mentioned, the Universal Hunab-Ku is beyond the range of our perception, but the galactic and earthly microcosms are within it, and each one of us is an individual expression of Hunab-Ku.

Our reception of new Heavens, new thoughts of God, propels us to participate in the construction of the Underworlds as agents of the divine plan. We may have been more like puppets of the process of creation than many would like to think. But spreading an awareness of how creation works and finding the path toward enlightenment may be

Figure 5.10. Hunab-Ku (literally, Hun Ahau, "one lord"), the Mayan yin/yang symbol for the Central Sun of the Universal World Tree, the One Source of Limits and Energy

just what is needed to liberate us from being such puppets and to expand the role of choice in our lives. In fact, if we want to climb to the next step of the pyramid, we need to assimilate this awareness. If it is our purpose to be created in the image of God, we cannot remain puppets. Being created in the image of God at this time, in the Galactic Underworld, means being creators in becoming; and understanding creation is one distinct step to be climbed on the cosmic pyramid. Taking responsibility as a cocreator with God presupposes a basic understanding of how creation works.

The holographic model of the universe that has been presented here now makes it possible for us to understand the phenomenon of *synchronicities,* a term introduced by the Swiss psychologist Carl Jung to describe statistically unlikely "chance" events that are experienced as meaningful. Because the Heavens, the thoughts of God, affect us all simultaneously, different individuals will come up with the same ideas independently at about the same time as a change takes place in the Heavens. With a cosmic projector automatically shifting archetypes influencing consciousness at calendar shifts, it would be much more surprising if ideas did *not* arise synchronistically in different places.

There are many other types of synchronicities as well. An example is when you run into a person just when you were thinking of him. Such an occurrence may be understood if the two persons were in resonance with the same Heaven. Chance meetings that are meant to happen occur because they are part of processes resulting from flows of divine light being focused in specific directions at calendar shifts. They are thus the results of the thoughts of God, and naturally many people use such synchronicities as a source of divine guidance. As the frequency of divine creation is now speeding up, such statistically impossible events have become commonplace, and at least the subconscious awareness is growing that there is a larger plan at work.

Yet few of us ever grow beyond the idea that synchronicities are strange and remarkable. This attitude is a logical consequence of the materialist worldview by which we are surrounded. This includes the denial by most people of a divine plan for the evolution of consciousness.

Because our frame of consciousness is to us like water is to a fish, we tend to deny the existence of consciousness and disregard the changes that it undergoes as the divine plan unfolds. This plan governs all aspects of our existence, including invisible changes in consciousness that generate unlikely coincidences, so it is to be expected that many events in the physical universe defy the laws of statistics. Yet blind to our own consciousness, we fall prey to the illusion that it is the material world that is real. With such an illusion, synchronicities must always appear strange and inexplicable. Developing intuition is about *tuning* in to the thoughts of God and looking through the veils of materialism, allowing the synchronicities to play for us.

BROADENING THE SCOPE OF HUMAN LIFE: THE CASE OF TELECOMMUNICATIONS

After these brief comments about the lower Underworlds, we may continue our climb of the Cosmic Pyramid. The Underworld that was created on top of the $13 \times 400 = 5,200$ tuns long National Underworld was the Planetary Underworld. Because the creation of each new Underworld is accomplished twenty times more rapidly than the creation of that below, the creation of the Planetary Underworld is accomplished in $13 \times 20 = 260$ tuns, or, in other words, 256.4 physical earth years. This Underworld began its rule on July 24, 1755, and its respective Days and Nights are katun periods of 20 tuns each, or 19.7 years (fig. 5.11).

The beginning of this Planetary Underworld in the mid-eighteenth century was also a time when very great changes occurred. A new age, a new civilization—the civilization of planetary industrialism—was seeded. In Europe its beginning saw the emergence of the Enlightenment and of the first cosmopolitans—persons who did not identify exclusively with their own nations but with humanity as a whole. The first Day of the Planetary Underworld also saw the beginning of the politically and philosophically important American Revolution. Most important, however, this was the beginning of the Industrial Revolution, which

Figure 5.11. The Thirteen Heavens of the Planetary Underworld with their ruling deities. The gods of light are shown on top and the gods of darkness underneath with the durations of the corresponding katuns.

is usually dated to James Watt's invention of the steam engine in 1769. The Planetary Underworld later developed many technical inventions that until recently few would have thought of as results of a cosmic plan.

The present book only gives a few examples of how the Mayan calendar describes evolutionary wave movements of consciousness: writing, religions, economics (appendix A), and now telecommunications. This brevity is not due to a scarcity of examples, but to the fact that the purpose of this book is to highlight the basic ideas rather than to describe the full versatility of the Mayan calendar. All things that evolve are strongly influenced by, and mostly directly subordinated to, the pattern of the seven Days and six Nights. Readers interested in seeing more examples are referred to my book *Solving the Greatest Mystery of Our Time: The Mayan Calendar* for a more complete picture.

A clear way of seeing the evolution, both of technology and of a planetary frame of consciousness, in the Planetary Underworld is to follow the development of telecommunications. When we discussed the National Underworld, I pointed out the important role that writing had in its development. As long as humans lived in a national context— which evolved at a reasonably slow baktun-based rhythm—communicating by written message was fast enough. Nations were usually so small that their entire areas could be reached by courier within a few days, and rarely was it necessary to communicate more rapidly than that. In the Planetary Underworld, however, dispatching papers was no longer a sufficiently swift means for communicating, as the rate of the process of creation, and hence that of technological development, sped up twentyfold to a katun-based rhythm. As a result of the evolution of the planetary frame of consciousness and its increased frequency, the whole planet has thus gradually developed into a common communication network. In this process, telecommunications—using the field of electricity that was essentially unknown in the National Underworld—has come to play the dominant role.

Figure 5.12 shows how the development of telecommunications has taken its greatest steps forward very close to the beginnings of the Days, which again are the odd-numbered Heavens, of the Planetary Underworld. We may also note how the potential for global contacts provided by these different types of telecommunication has expanded Day by Day. Step by step, the seed of the idea of a telegraph has developed into the fruit of the Internet, promoted by a wave movement propelling a new level of human creativity. Given the information in this table, we may wonder who is right. Is it the average person today who thinks that technological development is unguided, unplanned, and just the result of fortunate human inventiveness, or is it the Maya and the Mexica, who saw history as the result of creation cycles governed by different deities? It appears that the original Westerners have a strong case!

In figure 5.12 an evolutionary process is thus verified, which is a direct parallel to that described in figure 2.7 (page 28), although recreated at another level. Writing developed means of communication that were suitable within a nation, going from the first symbols etched into plates of clay to daily newspapers.

Telecommunications, in contrast, went from the first idea of the telegraph to the Internet. Both of these obviously resulted in an enormous increase in speed for communications, but the latter represents an evolution of means of communication that serve the integration of a whole planet. In this we may clearly see how the two Underworlds, the National and the Planetary, correspond to two different frames of consciousness. Yet they both follow the same basic pattern of an evolution from seed to mature fruit taking place in seven Days of creation.

Growth Stage Day and Heaven No. Ruling Deity	Time Span	Invention or Development
Sowing Day 1, Heaven 1 god of fire and time	1755–1775	**Theory of telegraph** Anonymous (1753) Bozolus (1767)
Germination Day 2, Heaven 3 goddess of water	1794–1814	**Optical telegraph** Chappe, Paris-Lille (1794) Sweden (1794)
Sprouting Day 3, Heaven 5 goddess of love and childbirth	1834–1854	**Electrical telegraph** Morse (1835) Washington-Baltimore line (1843)
Proliferation Day 4, Heaven 7 god of maize and sustenance	1873–1893	**Telephone** Bell's patent application (1876) First telephone station (in U.S., 1878)
Budding Day 5, Heaven 9 god of light	1913–1932	**Radio** First regular broadcast (in U.S., 1910; in Germany, 1913)
Flowering Day 6, Heaven 11 goddess of birth	1952–1972	**Television** First public broadcast (in U.K., 1936) First color TV broadcast (in U.S., 1954)
Fruition Day 7, Heaven 13 Dual-Creator God	1992–2011	**Computer networks** Internet (1992) Global television channels Mobile telephones

Figure 5.12. The evolution of telecommunications from the theory of the telegraph (seed) to cell phones and the Internet (fruits) in the Days of the Planetary Underworld

As shown in figure 5.11, it is the same thirteen deities, reflecting the same energies, that rule the sequence of thirteen katuns in the Planetary Underworld as those that ruled the thirteen baktuns of the National Underworld (see fig. 2.5, page 24). This pattern, which is common to all Underworlds, is the very basis of Mayan calendar prophecy. Because of the recurrence of the same sequence of deities—and energies—in different Underworlds, it is possible to make predictions about the future, something we will see more of later.

THE FREQUENCY INCREASE AND
THE SPEEDING UP OF TIME

The World Tree is the primary generator of shifts between Days and Nights. We may then realize that the frequency of such shifts increases twentyfold with every higher Underworld. Our experience of time is thus associated with a specific frequency at each level of consciousness of the cosmic pyramid. The speeds of the different means of communication (see fig. 2.7, page 28, and fig. 5.12, page 109), for instance, are simply manifestations of these frequencies. Each Underworld is thus dominated by a basic frequency for the evolution of consciousness, and these frequencies increase from the very lowest of 1/hablatun to the very highest of 1/uinal as the nine-story pyramid is climbed. This 25 billion-fold increase in the basic tones of creation (there surely are a number of overtones too) affects the type of phenomena generated by the various Underworlds. Thus the Invisible Cross may be seen as an instrument programmed to generate a cosmic symphony, where tones of higher and higher frequencies come into play in a preset pattern. (A better metaphor than a symphony would be a conductor starting by cuing the drums, and then cuing one instrument after another to add to those that are already playing.)

This pyramidal structure of the frequencies of creation is behind the common experience that time is now going much faster—to a point where it sometimes seems to disappear. Every level, every terrace of the pyramid, corresponds to a certain level of consciousness, but it obviously also corresponds to a certain frequency of change (see fig. 5.5, page 94). With every higher level of the pyramid of cosmic creation, a twentyfold acceleration of spiritual time takes place. The more frequent the alternations between Days and Nights, the higher the frequency of change and the more rapidly human beings experience the passage of time. Thus, for instance, in the Galactic Underworld, as much change must happen in a tun (360 days) as happened in a katun (19.7 years) during the Planetary Underworld, or in a baktun (394 years) of the National Underworld.

Today we should be aware of this origin of the acceleration of time. The so-called burnout phenomenon has increased dramatically in many countries where people are victimized by stress, seemingly because of the high pace of work around them. In reality burnout is a result of the many increasingly more pronounced conflicts between the consciousness generated by the materialist Planetary Underworld and that being created by the newly emerging, more spiritual Galactic Underworld. As a result, the symptoms of burnout are also becoming more serious. The only true remedy is to align our lives with the cosmic evolution toward enlightenment and to focus our whole existence—thinking, acting, and being—on this. In the cosmic plan there is no turning around. The values of the Planetary Underworld, one way or another, must yield to those of the Galactic.

Ultimately, the phenomenon of stress arises from the conflict in the individual between cosmic, divine time and physical time. Many have found that using an ordinary wristwatch is an impediment to going with the flow of time; without this device, synchronicities occur more easily. Divine time is the flow of the Galactic Underworld leading toward enlightenment according to the tun-based Mayan calendar, whereas physical, measured time is subordinated to physical processes. The watch is a mechanized tool for measuring time, and it subordinates us to a whole set of rules for life typical of the industrialized world of the Planetary Underworld. When the individual wants to go with the flow of divine time, while the watch—based on physical time—tells him or her to do otherwise, a stressful conflict arises.

In principle, however, this conflict is not any different from that between, on the one hand, following the solar year, regardless of how it is sliced into months, and, on the other hand, following the Mayan calendar. The mechanical movements of the watch are directly linked to the mechanical movements of the astronomical cycles, and these likewise tend to subordinate human beings to the physical reality rather than creating a mind-set that supports them in going with the flow of divine time. The liberation of human beings from the negative effects of the frequency increase, such as stress, is a matter of entering the natural, nonphysical, divine flow of creation leading toward enlightenment. Today more than ever before, people are becoming aware of the stressful effects of subordinating their lives to physical time.

Calendars based on physical time, such as the Gregorian, Muslim, or Jewish calendars, or common astrology for that matter, cannot explain the current acceleration of time and the increase of stress. The duration of the solar year has not changed much in the past million years. Only the hierarchical, tun-based Mayan calendar system, which is based on the oscillations of the World Tree rather than movements of material bodies, is able to explain the acceleration of time. With this calendrical system the increasing pace of time becomes a logical consequence.

Ultimately, the increase in frequency carried by the only thirteen-tun-long Galactic Underworld is behind the current revival of the tzolkin and the Mayan calendar. Only as the changes between Days and Nights occur at a very high frequency do people on a larger scale become aware that there are spiritual cosmic factors behind the changing tides of human creativity. As these changes are now happening so frequently—every 360 days—it seems that many, at least on a subconscious level, are becoming aware of them. This increase in frequency will become progressively more evident as the Galactic Underworld becomes more dominating, and even more so as the Universal Underworld draws near. But more on this later.

THE PLANETARY ROUND OF LIGHT AND THE ASCENSION OF THE COSMIC PYRAMID

The nine different Underworlds that form the cosmic pyramid of consciousness have now been presented. Nothing has yet implied, however, that this cosmic pyramid can lead humanity to the state of enlightenment. There is one more very important principle to be introduced into the overall picture of the calendrical system of the Maya: the Planetary Round of Light.

This principle says that each of the Nine Underworlds is dominated by a specific yin/yang (darkness/light) polarity and that these polarities vary among Underworlds according to a certain pattern. The cosmic plan must include such a pattern to be able to lead humanity on the path toward enlightenment. The underlying reason for the existence of these shifting polarities is obviously that if all Nine Underworlds favored a specific hemisphere, events in the world would have gone out of balance a long time ago. To counteract this, the Planetary Round of Light serves to balance the effects of the various Underworlds. This is the fundamental reason that even if a Fall is still affecting us, there is also a possible path out of this to a glorious future, as has been promised in many religious and spiritual traditions.

What all previous attempts to understand the Mayan calendar have had in common is a failure to recognize that this calendar system describes the evolution of consciousness in nine different Underworlds. Through comparisons among the energies of the Thirteen Heavens in these Underworlds, we can learn something about these energies; it is only the Nine Underworlds taken together that make the Mayan calendar a prophetic guide on the path toward enlightenment. The Mayan calendar encompasses all of creation, built by Nine Underworlds. The most significant pyramids in the Mayan area still stand as evidence of the prominence of these Underworlds. To understand the fundamental message of these pyramids, all we really need do is look at what is immediately obvious: they have nine levels.

The yin/yang polarities of the Planetary Round of Light are pictured in figure 5.13, where the earth is shown from the "back side" (that is, the "side" opposite the Pacific, which is in resonance with the eyes and the face) and from above the North Pole, respectively. As the nine-story pyramid is climbed, the yin/yang polarity turns 90° counterclockwise with every new level of consciousness. As the World Tree, the polar axis, exercises its effects through oscillating pulses in perpendicular directions (one of which is longitude 12° East), it generates ethereal creation fields with alternating yin/yang polarities. As discussed previously, the model presented here is consistent with the hypothesis that the earth's inner core encompasses an octahedral

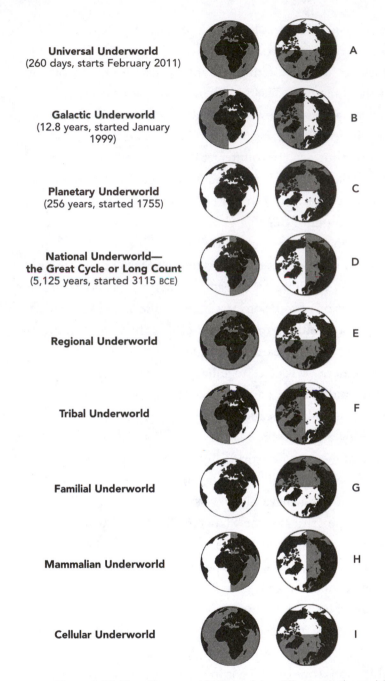

Universal Underworld
(260 days, starts February 2011)

A

Galactic Underworld
(12.8 years, started January 1999)

B

Planetary Underworld
(256 years, started 1755)

C

**National Underworld—
the Great Cycle or Long Count**
(5,125 years, started 3115 BCE)

D

Regional Underworld

E

Tribal Underworld

F

Familial Underworld

G

Mammalian Underworld

H

Cellular Underworld

I

Figure 5.13. The yin-yang polarities of the different Underworlds in the Planetary Round of Light. With every new higher level of the cosmic pyramid, the polarities shift 90° as seen from the polar axis. On the left, the earth is shown from the "back side," and on the right it is shown from above the North Pole.

Figure 5.14. Quetzalcoatl ascending a five-story pyramid, which may signify the climb up and down the Nine Underworlds. In the East sacred buildings are, like this pyramid, built in five stories. The ahauob, the Mayan shaman-kings, would climb to the top of the nine-story pyramids to perform their ceremonies. By doing so they sought, at least symbolically or in a trance state, to ascend to the universal level of consciousness and so return to unity with the cosmos.

structure. Thus with every new Underworld a new yin/yang polarity is added to the already existing ones as the line of separation generated by the World Tree shifts 90°.

Because the brains and minds of human beings are in holographic resonance with the earth, as the cosmic pyramid is ascended, their resulting frames of consciousness will be dominated by the corresponding yin/yang polarities.

Some of these yin/yang polarities create a separation between the functioning of the brain hemispheres of human beings and so create a dualist mind (reality is perceived through a dualist filter), whereas other polarities create a unitary mind (reality is perceived as unitary). The dualist Underworlds may filter out either the aspects of reality perceived by the Western/left-brain hemisphere or those perceived by the Eastern/right-brain hemisphere. When it comes to the unitary Underworlds, some of these blindfold human beings, whereas others make them enlightened. As with many other assertions in this book, however, there is not space to present the empirical evidence for this Round of Light, so the reader is simply invited to try on this perspective and see if it fits his or her intuition.

To make a brief survey of the Planetary Round of Light, let us begin with the unitary consciousness of the Regional Underworld. (In this survey it will serve to imagine your head surrounded by the globe, with your eyes looking out through the Pacific Ocean, the "face" of the planet. As the yin/yang polarities shift with the different Underworlds, reality will be perceived through different filters.) The Regional Underworld roughly corresponds to early cultured humans, who during the Ice Age herded reindeer, painted cave art, and made some complex tools. Human beings presumably lived in egalitarian communities with strong matriarchal features. In this Underworld they were enlightened, seeing no separation between themselves and the divine, living cosmos. Nor was there a separation between man and woman or between humans and nature. This level of consciousness was symbolically described as the Garden of Eden. In it humans lived fully in the present and were not concerned about past or future. There were no pharaohs and no slaves, but also no irrigation systems and certainly no conveniences. This was an enlightened state within a very limited frame of consciousness. Through the Planetary Round of Light, however, divine providence had greater plans for the destiny of humankind.

The beginning of the National Underworld amounted to the Fall and the emergence of a dualist consciousness that laid the foundation for the modern mind. The human qualities associated with the right-brain hemisphere were filtered out and repressed, and the modern mind was created. People came to be valued differently; women, children, and manual workers, as well as animals, were downgraded and relegated to the yin (darkness) field of perception in this frame of consciousness. Yet humans were still "semi-enlightened," and as their inner vision could see at least half the light, they retained a belief in a spiritual reality. Nevertheless, the living contact with the divine was gradually lost, and spirituality turned into religion. More and more the Western/left-brain hemisphere was favored, as were its expressions—science, Protestant Christianity, and capitalism.

With the next level, the Planetary Underworld, humans acquired, unbeknownst to themselves, a consciousness that was at the same time unitary and blindfolded. Because it was blindfolded, humans could no longer see divine reality, and as a result strong agnostic and atheist tendencies developed. Human existence focused on production and consumption and struggles over the material gains of industry. Within this frame of consciousness only the material aspects of reality were visible to humans; if religion did not disappear altogether it was reduced to little more than formal ritual. The frame of consciousness carried by this Underworld, however, was also unitary, and so some of the most extreme expressions of the tendency to make distinctions between people began to moderate. (The basic hierarchical structure remained, but a measure of equality emerged, because the blindfolded consciousness no longer projected inequality.) Thus in the mid-eighteenth century, women first emerged as a political force and the special needs of children were acknowledged. Democracy emerged, slavery was outlawed in Europe, and power and wealth based on inherited privilege began to disappear. For the first time in human history peace, although rarely practiced, became an ideal.

With the recent beginning of the Galactic Underworld, humankind is now gradually acquiring a frame of consciousness that favors the Eastern/right-brain hemisphere. We can expect that this will create a balance between East and West, as well as between the right and left-brain hemispheres. The four-century dominance of the world by the Western/left-brain hemisphere brought about by the Thirteenth Heaven of the National Underworld will come to an end. Yet since the emerging Underworld carries a dualist consciousness, there is no reason to be surprised that it has initially caused a dramatic increase in hostilities between East and West. This is also a "semi-enlightened" frame of consciousness, meaning that the spiritual aspects of reality will once again be recognized. The imposition of this frame means a change that will affect all aspects of our existence. Women, who have somewhat better communication between their brain hemispheres (which is to some extent shared by homosexual men), will in many ways be the first to integrate these new ways of perceiving reality.

With the Universal Underworld, humankind will finally attain a consciousness that is both unitary and enlightened, based on the balance created by the lower Underworlds (if you superimpose the four yin/yang polarities of the Regional through Galactic Underworlds on one another, they balance out). The Universal Underworld means adding one enlightening yin/yang polarity on top of all the lower frames that have already been evened out. This will mean a lasting peace, both on the outer and the inner planes. It will also mean an end to the separation between humans and the divine and the attainment of an enlightened state that cannot be reversed. This is the paradise at the end of time that has been prophesied in many religions. Because the World Tree, the wave generator of energies, will no longer create a perceived division line between the hemispheres, history—as well as our experience of time itself—will come to an end.

What it means to be a human being differs widely from one Underworld to another, because human nature is profoundly influenced by their alternating cosmic energies and yin/yang polarities. These yin/yang polarities determine our frames of consciousness that provide the very filters through which we perceive reality. To talk about human nature as something fixed truly lacks a basis in reality. Nor can human evolution be explained simply by superficial changes in thinking. Differences in consciousness between people living in different Underworlds are real, and this is also why we should avoid being judgmental about history. The shifting of yin/yang polarities explains why the ancients could see things that we can no longer see, and explains why we now have something to learn from the Mayan calendar. (There are also some things that modern people can see that the ancients could not, so we should be careful not to throw the baby out with the bathwater!)

As we climb to a higher Underworld on the cosmic pyramid, our frame of consciousness is transformed. The path toward enlightenment is, then, the transformation entailed by the climb to the highest consciousness of the Universal Underworld. The Mayan calendar system allows us to give a precise meaning to enlightenment, which is a timeless cosmic consciousness in which our experience of reality is not dominated by a yin/yang polarity that generates outer or inner conflicts. It is the purpose of the Universal Underworld, and thus of this whole creation, to develop an enlightened consciousness. For this transformation process the Mayan calendar is an invaluable map of time describing the changing spiritual terrains and perceptions of reality, a map of our ascension to the highest level of consciousness. Those who have set enlightenment as the focus of their path will find this map invaluable.

In Native American legends, yin/yang polarities were commonly symbolized by pairs of brothers or twins. For instance, the twins Hunahpu/Xbalanque among the Maya, and Quetzalcoatl/Tezcatlipoca among the Aztecs, symbolized the polarity between yang and yin. Similarly, in the Hopi Prophecy it is said that the Elder Brother of Light, the True White Brother, will return from the East to help his younger brother purify the West. This sounds exactly like a description of the particular yin/yang polarity that is now emerging with the Galactic Underworld. This new light from the East would lead the world away from the present state of *Koyaanisqatsi,* imbalance, to one nation under one power, the Creator. To signal the change, a dancing kachina will remove his mask, signifying the removal of darkness from the face of the world—a symbol of what happens as the enlightened state of the Universal Underworld emerges.

6

.

The Tzolkin

AS IN THE LONG CYCLES, SO IN THE SHORT

After this brief outline of the longer waves of creation and the climb to enlightenment generated by the alternating yin/yang polarities of the cosmic pyramid, we return to the study of the tzolkin. This Sacred Calendar is both a master calendar providing a unifying pattern for all creation and the hub of the Mayan calendar system. Those seeking to understand the current and future course of events should familiarize themselves with its basic pattern. Although references to the tzolkin have so far been made only at intervals, this is in fact the structure underlying the prophetic content of the Mayan calendar system.

Figure 4.5 (page 78), where the seven Days and six Nights of the Great Cycle were applied to the 13 × 20 pattern of the tzolkin, indicated how the larger creation cycles are related to the tzolkin chart. The tzolkin is a matrix for the energies of the divine process of creation. Crucial for our understanding of the day-to-day tzolkin is what may be called the Hermetic principle of time: "As in the long cycles, so in the short." This principle means that the basic energy pattern of seven Days and six

Tzolkin Chart 4 with Uinals

Mayan Day Signs   ## Aztec Day Signs

Mayan														Aztec
Imix	1	21	41	61	81	101	121	141	161	181	201	221	241	Cipactli
Ik	2	22	42	62	82	102	122	142	162	182	202	222	242	Ehecatl
Akbal	3	23	43	63	83	103	123	143	163	183	203	223	243	Calli
Kan	4	24	44	64	84	104	124	144	164	184	204	224	244	Cuetzpallin
Chicchan	5	25	45	65	85	105	125	145	165	185	205	225	245	Coatl
Cimi	6	26	46	66	86	106	126	146	166	186	206	226	246	Miquiztli
Manik	7	27	47	67	87	107	127	147	167	187	207	227	247	Mazatl
Lamat	8	28	48	68	88	108	128	148	168	188	208	228	248	Tochtli
Muluc	9	29	49	69	89	109	129	149	169	189	209	229	249	Atl
Oc	10	30	50	70	90	110	130	150	170	190	210	230	250	Itzcuintli
Chuen	11	31	51	71	91	111	131	151	171	191	211	231	251	Ozomatli
Eb	12	32	52	72	92	112	132	152	172	192	212	232	252	Malinalli
Ben	13	33	53	73	93	113	133	153	173	193	213	233	253	Acatl
Ix	14	34	54	74	94	114	134	154	174	194	214	234	254	Ocelotl
Men	15	35	55	75	95	115	135	155	175	195	215	235	255	Cuauhtli
Cib	16	36	56	76	96	116	136	156	176	196	216	236	256	Cozcacuauh
Caban	17	37	57	77	97	117	137	157	177	197	217	237	257	Ollin
Etznab	18	38	58	78	98	118	138	158	178	198	218	238	258	Tecpatl
Cauac	19	39	59	79	99	119	139	159	179	199	219	239	259	Quiahuitl
Ahau	20	40	60	80	100	120	140	160	180	200	220	240	260	Xochitl

Figure 6.1. Tzolkin chart with seven uinals of light and six uinals of darkness. This is the pattern applied to the National Underworld in figure 4.5 (page 78).

Nights that can be empirically verified in all Nine Underworlds is also reflected in the daily tzolkin.

As we shall see, the pattern of seven Days and six Nights in figure 6.1 is just the most basic aspect of the tzolkin pattern. This basic subpattern generates the wave movement for the evolution of consciousness in thirteen steps, as shown in figure 6.2. This pattern applies to all evolutionary processes, regardless of which Underworlds developed them. Hence the microcosmos reflects the macrocosmos in the dimension of time as well.

The common 260-day tzolkin round is a temporal microcosm of the longer creation cycles, in which the pattern of energy changes of the tzolkin rounds is reflected. In reality the light pattern of the tzolkin, which may be applied to different sequences of 260 consecutive equal time periods in the tun-based system, defines the process of creation in all the Underworlds. The tzolkin pattern is thus intimately linked to the unfolding of a frame of consciousness, and a given tzolkin energy always represents a specific point in its evolution. The tzolkin is really a filtration pattern of divine light. This pattern exists in a realm beyond time, in the mind of God, as it were. The tzolkin by itself is thus a pattern for the unfolding of creative energy rather than a calendar.

As discussed earlier, the tzolkin may also be visualized as two interconnected cogwheels (see fig. 1.8, page 13). Similarly, a way of looking at creation in its entirety is to visualize it as a set of interlocking cogwheels that are in phase with each other (fig. 6.3). These cogwheels fit with each other only according to certain rules, however. Each of the eighteen uinals of the tun must, for instance, be exactly matched with each of the thirteen uinals that make up a tzolkin round. Because each wheel in the tun-based system (tuns, katuns, baktuns, and so on) has twenty times more cogs than that on the inside, an inner wheel will rotate twenty times in the same amount of time as an outer wheel rotates once. In this way it becomes evident that each cycle of time has its own rotation frequency, and that this frequency varies among different Underworlds. Thus all cycles of spiritual time in the universe are interconnected. Nothing happens in isolation from the large-scale cosmic plan, and everything is subjected to the same cycles of time.

Heaven and Ruling Aztec Deity	Day/Night	Phase in the Development of Consciousness or Phenomenon
First Heaven **Xiuhtecuhtli,** god of fire and time	Day 1	**Sowing** Initial expressions of new consciousness
Second Heaven **Tlaltecuhtli,** god of earth	Night 1	Inner assimilation of new consciousness
Third Heaven **Chalchiuhtlicue,** goddess of water	Day 2	**Germination** Expansive push of new consciousness
Fourth Heaven **Tonatiuh,** god of the sun and warriors	Night 2	Reaction against new consciousness
Fifth Heaven **Tlacolteotl,** goddess of love and childbirth	Day 3	**Sprouting** Establishment of new consciousness
Sixth Heaven **Mictlantecuhtli,** god of death	Night 3	Inner assimilation of established new consciousness
Seventh Heaven **Cinteotl,** god of maize and sustenance	Day 4	**Proliferation** New consciousness balances old consciousness Most pertinent expression of new consciousness
Eighth Heaven **Tlaloc,** god of rain and war	Night 4	Bridge-building to prepare for expansion of new consciousness
Ninth Heaven **Quetzalcoatl,** god of light	Day 5	**Budding** Breakthrough of most pertinent expression
Tenth Heaven **Tezcatlipoca,** god of darkness	Night 5	Destruction
Eleventh Heaven **Youhalticitl,** goddes of birth	Day 6	**Flowering** Renaissance, protoform of highest expression of new consciousness
Twelfth Heaven **Tlahuizcalpantecuhtli,** god before dawn	Night 6	Resting and fine-tuning of protoform of new consciousness
Thirteenth Heaven **Ometeotl/Omecinatl,** Dual-Creator God	Day 7	**Fruition** Manifestation of highest expression of new consciousness

Figure 6.2. The phases in the evolution of a phenomenon that passes through the energies of the Thirteen Heavens. This evolutionary progression is applicable to any of the Nine Underworlds.

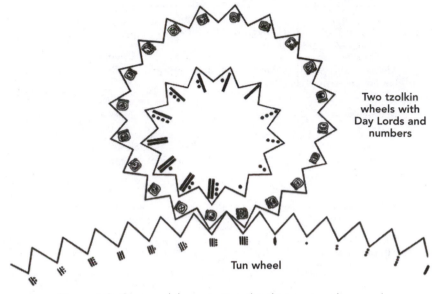

Two tzolkin wheels with Day Lords and numbers

Tun wheel

Figure 6.3. Cog model connecting the thirteen numbers and the twenty day signs with a tun wheel

It now also becomes apparent that the Mayan calendar system is one of cycles within cycles within cycles, where the more detailed day-to-day course of events is conditioned by the energies of the shorter cycles. In divine creation everything is related, and the various calendrical cycles describing its evolution are related in clearly defined ways.

This also means that the shorter periods, such as the twenty-day uinals or the thirteen-day trecenas of the tzolkin, are high-frequency overtones of the tones generated by the Great Cycle and other Underworlds of even longer duration (as in the long cycles, so in the short). Because we know from human history that the wave movements developing these longer creation cycles are real, the microcosmic reflections of these cycles must also be real. The same high-frequency energy pattern of the tzolkin that dominates the longer cycles is recreated in the 260-day cycles. As a consequence, there is a real basis for Mayan "astrology" and divination, as is discussed in appendix B.

The nature of the tzolkin as primarily a spiritual energy pattern is also evident from the fact that among both the Maya and the Mexica, gods were often symbolized by, and talked about as, tzolkin combinations. Quetzalcoatl, for example, was 9 Wind (Bolon Ik). Although the tzolkin was initially described as a count of 13 × 20 regular days—which is the way the tzolkin is primarily known—such a view is really very constrained. In the ancient Books of Chilam Balam, the 260 combinations of the tzolkin are sometimes linked to katuns, whose progression describes a level of the creation of the universe. The model developed in this book is completely in line with this. The Creator seems to have concluded that the tzolkin is the optimal pattern of creation, the pattern behind the evolution of all things. This realization of the Maya and the Mexica, that the tzolkin is a filtration pattern of light that pervades the universe and seeks expressions on many different levels, is one of their greatest contributions to human spirituality.

Ultimately, the tzolkin is a code that describes a pre-set divine program for the oscillations of the World Tree, a program that was conceived beyond space and time. The tzolkin is much larger and much more profound than any lunar cycle, biorhythm, or solar phase. On a deeper level the tzolkin is timeless. If its changing energies create our experience of time, it should also be obvious that time is an illusion and has no existence independently of divine creation. The tzolkin is, more than anything else, the pattern ultimately behind all rhythms and all structuring of energies.

THE TRUE TZOLKIN COUNT

The cogwheel model (fig. 6.3, page 124) has one very important consequence: There can only be one true tzolkin count as far as the cosmos goes. In a sequence of thirteen days, for example, there can be only one of the days that is in resonance with the Eighth Heaven of all the different Underworlds (ruled by the energy of Tlaloc in all of them). This is the day given the number 8 out of the thirteen numbers of the tzolkin count. A certain day is thus dominated by one and only one tzolkin combination. Overtones must have harmonious relationships to tones. If a baktun, a katun, or a tun did not end at exactly the same day as a uinal of the tzolkin, the tzolkin rounds would not be exact microcosmic reflections of the energies of the longer cycles. Moreover, the tzolkin and the tun-based system must always be in phase.

The classical count that has been in use among the Maya for some 2,500 years is harmoniously related to the larger cycles. (The first days, 1 Imix, *kin* 1, of the tzolkin rounds of this count can be seen in the middle row of fig. 7.1, page 138). Over time, however, a few minor groups of Maya have for a variety of reasons lost the classical count and replaced it with new ones. (The tzolkin count still in use among the Maya is usually referred to as 584, 283 count, a number that sometimes needs to be plugged in to Mayan calendar calculators on the Internet.)

Also creating confusion are a number of invented tzolkin counts, notably the so-called Dreamspell, which are widespread on the Internet. The Dreamspell count was invented some ten years ago and has never been used by the Maya. A crucial characteristic of this count is that it has a leap day—February 29 as defined by the Gregorian calendar—every four years that lacks a tzolkin energy. To have such a day that is not ruled by a day sign (as if creation somehow pauses then) is very alien to Mayan day keepers. Because the Dreamspell count jumps over days, it does not have a stable relationship to the tun-based system of the Nine Underworlds, including the Thirteen Heavens of the Great Cycle. Hence it is not an adequate reflection of the evolution of consciousness such as has been described in this book.

The tzolkin is usually described as the central hub of creation. This may be why the Maya have placed so much emphasis on it and have kept it intact and unchanged to this day. To change the tzolkin may create much confusion, and it is very likely to lead to a materialist notion of time. At the very least, someone who uses an inaccurate tzolkin count cannot determine her or his true tzolkin energy of birth and learn to flow with the divine process of creation.

SUBDIVIDING THE TZOLKIN: THE LAYERING OF FILTRATION PATTERNS OF DIVINE LIGHT

The meaning of the cycle of thirteen numbers has been fairly extensively discussed, and a few examples have been given as to how these correspond to an evolutionary progression from seed to mature fruit. The seven uinals of light and the six uinals of darkness, however, are only the very simplest way in which creative energy is divided by the tzolkin pattern (fig. 6.1, page 121).

The tzolkin may also be divided into several other important subpatterns, and its 260 units may be factored in the following ways: 2×130, 4×65, 5×52, 10×26, 13×20, 20×13, 26×10, 52×5, 65×4, and 130×2. Each such factoring creates a subpattern of divine light of its own, so the complete energy matrix of the tzolkin emerges only as all these subpatterns are combined. The surviving Mayan books, and especially the Dresden Codex, deal extensively with how such subpatterns of the tzolkin are generated, and the same can be said about some of the preconquest Mexica codices, such as the Codex Borgia. Only as all of these are layered one upon another does the complete light filtration pattern of the tzolkin emerge.

In figure 6.4 we see how the basic filtration pattern of seven uinals of light and six of darkness is combined with a subdivision into twenty trecenas. The energy of each trecena is believed to be dominated by its initial day sign, so these have been introduced into this tzolkin pattern.

The tzolkin pattern in figure 6.5 is even more advanced and detailed. In this the energies shifting between Days and Nights within each trecena have been overlaid, so this is even closer to the truth than those in figures 6.1 and 6.4. The next section discuss two more overlying patterns, those of four and five Worlds.

It should be obvious that the tzolkin, and divine creation, may be studied at any level of complexity. In fact, to be properly understood, many evolving phenomena in cosmic history require a more complex tzolkin pattern than that of seven Days and six Nights. Interestingly, it can be shown mathematically that if all the subpatterns of the different factorizations are layered onto the tzolkin, each of the 260 units will have its own unique energy.

Tzolkin Chart 4
with Uinals and Trecenas

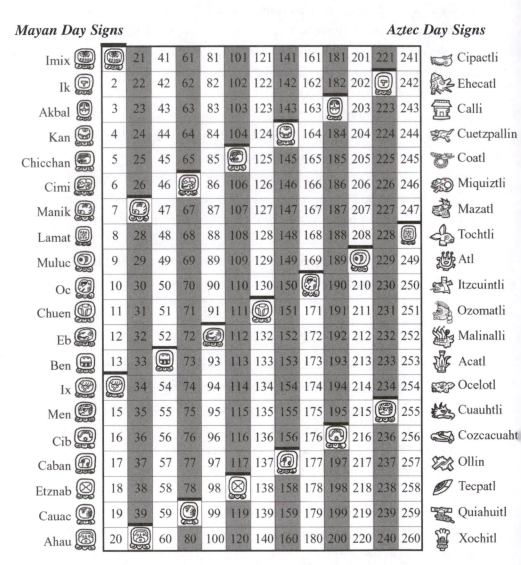

Mayan Day Signs													Aztec Day Signs
Imix		21	41	61	81	101	121	141	161	181	201	221	241	Cipactli
Ik	2	22	42	62	82	102	122	142	162	182	202		242	Ehecatl
Akbal	3	23	43	63	83	103	123	143	163		203	223	243	Calli
Kan	4	24	44	64	84	104	124		164	184	204	224	244	Cuetzpallin
Chicchan	5	25	45	65	85		125	145	165	185	205	225	245	Coatl
Cimi	6	26	46		86	106	126	146	166	186	206	226	246	Miquiztli
Manik	7		47	67	87	107	127	147	167	187	207	227	247	Mazatl
Lamat	8	28	48	68	88	108	128	148	168	188	208	228		Tochtli
Muluc	9	29	49	69	89	109	129	149	169	189		229	249	Atl
Oc	10	30	50	70	90	110	130	150		190	210	230	250	Itzcuintli
Chuen	11	31	51	71	91	111		151	171	191	211	231	251	Ozomatli
Eb	12	32	52	72		112	132	152	172	192	212	232	252	Malinalli
Ben	13	33		73	93	113	133	153	173	193	213	233	253	Acatl
Ix		34	54	74	94	114	134	154	174	194	214	234	254	Ocelotl
Men	15	35	55	75	95	115	135	155	175	195	215		255	Cuauhtli
Cib	16	36	56	76	96	116	136	156	176		216	236	256	Cozcacuauht
Caban	17	37	57	77	97	117	137		177	197	217	237	257	Ollin
Etznab	18	38	58	78	98		138	158	178	198	218	238	258	Tecpatl
Cauac	19	39	59		99	119	139	159	179	199	219	239	259	Quiahuitl
Ahau	20		60	80	100	120	140	160	180	200	220	240	260	Xochitl

Figure 6.4. Tzolkin chart with two overlaid patterns: the pattern of seven uinals of light and six of darkness, and the pattern of the twenty trecenas, each marked by its initiating sign and separated by a thick bar

FOUR AND FIVE WORLDS

Among the possible factorizations of the tzolkin, those of 4 × 65 (fig. 6.6) and 5 × 52 (fig. 6.7) units may be the most prominent. These factorizations generate four (or five) different Worlds that are commonly part of Mesoamerican and Native American prophetic traditions (although many such prophecies have lost their direct connection to the tzolkin). In the codices it seems that a tzolkin divided into four equal segments (4 × 65 = 260), resulting in four Worlds (fig. 6.7), played a very important role in ancient times. Because of the importance of the energy shifts between these four Worlds, their respective time spans are given in four different Underworlds in figure 6.8.

In figure 6.8, note that in the National Underworld the third of these four Worlds began at its very midpoint, in 552 B.C.E. As mentioned, this time was marked by an exceptionally strong burst of religious thought worldwide (see fig. 4.6, page 81). In contrast, in the fourth of these Worlds, beginning in 730 C.E., a consciousness with an opposite effect was implemented (see fig. 5.13, page 113) that began to prepare humanity for the Planetary Underworld with its materialism and industrialism that truly began to manifest only in 1755.

THE AZTEC CALENDAR STONE

The cosmology of the Mexicas was in significant ways built on the tzolkin. This is very clear from their Calendar Stone, which is on display at the Anthropological Museum in Mexico City. The stone itself is one of the most famous symbols of the nation of Mexico, and much research has been undertaken to understand the meaning of its symbols in the different rings surrounding the face of Tonatiuh, the god of the sun and warriors. Tourists in Mexico are literally bombarded by souvenirs of this Calendar Stone, which is sometimes confused with the Mayan calendar.

The Calendar Stone (fig. 6.9) is a description of the cosmology of the Mexicas, who had a World-based creation scheme in which four

Tzolkin Chart 5
with Uinals, Trecenas, Days and Nights

Mayan Day Signs · *Aztec Day Signs*

Mayan Day Sign														Aztec Day Sign
Alligator	1	21	41	61	81	101	121	141	161	181	201	221	241	Alligator
Wind	2	22	42	62	82	102	122	142	162	182	202	222	242	Wind
Night	3	23	43	63	83	103	123	143	163	183	203	223	243	House
Seed	4	24	44	64	84	104	124	144	164	184	204	224	244	Lizard
Serpent	5	25	45	65	85	105	125	145	165	185	205	225	245	Serpent
Death	6	26	46	66	86	106	126	146	166	186	206	226	246	Death
Deer	7	27	47	67	87	107	127	147	167	187	207	227	247	Deer
Rabbit	8	28	48	68	88	108	128	148	168	188	208	228	248	Rabbit
Water	9	29	49	69	89	109	129	149	169	189	209	229	249	Water
Dog	10	30	50	70	90	110	130	150	170	190	210	230	250	Dog
Monkey	11	31	51	71	91	111	131	151	171	191	211	231	251	Monkey
The Road	12	32	52	72	92	112	132	152	172	192	212	232	252	Grass
Reed	13	33	53	73	93	113	133	153	173	193	213	233	253	Reed
Jaguar	14	34	54	74	94	114	134	154	174	194	214	234	254	Ocelot
Eagle	15	35	55	75	95	115	135	155	175	195	215	235	255	Eagle
Vulture/Owl	16	36	56	76	96	116	136	156	176	196	216	236	256	Vulture
Earth	17	37	57	77	97	117	137	157	177	197	217	237	257	Movement
Flint	18	38	58	78	98	118	138	158	178	198	218	238	258	Knife
Rainstorm	19	39	59	79	99	119	139	159	179	199	219	239	259	Rain
Light/Lord	20	40	60	80	100	120	140	160	180	200	220	240	260	Flower

Figure 6.5. Tzolkin chart with overlaid patterns like those in figure 6.4, plus an overlay of the daily changes between light and darkness within the trecenas

Tzolkin Chart 6
with Four Worlds

Mayan Day Signs*   *Aztec Day Signs*

Mayan														Aztec
Imix	1	21	41	61	81	101	121	141	161	181	201	221	241	Cipactli
Ik	2	22	42	62	82	102	122	142	162	182	202	222	242	Ehecatl
Akbal	3	23	43	63	83	103	123	143	163	183	203	223	243	Calli
Kan	4	24	44	64	84	104	124	144	164	184	204	224	244	Cuetzpallin
Chicchan	5	25	45	65	85	105	125	145	165	185	205	225	245	Coatl
Cimi	6	26	46	66	86	106	126	146	166	186	206	226	246	Miquiztli
Manik	7	27	47	67	87	107	127	147	167	187	207	227	247	Mazatl
Lamat	8	28	48	68	88	108	128	148	168	188	208	228	248	Tochtli
Muluc	9	29	49	69	89	109	129	149	169	189	209	229	249	Atl
Oc	10	30	50	70	90	110	130	150	170	190	210	230	250	Itzcuintli
Chuen	11	31	51	71	91	111	131	151	171	191	211	231	251	Ozomatli
Eb	12	32	52	72	92	112	132	152	172	192	212	232	252	Malinalli
Ben	13	33	53	73	93	113	133	153	173	193	213	233	253	Acatl
Ix	14	34	54	74	94	114	134	154	174	194	214	234	254	Ocelotl
Men	15	35	55	75	95	115	135	155	175	195	215	235	255	Cuauhtli
Cib	16	36	56	76	96	116	136	156	176	196	216	236	256	Cozcacuauhtli
Caban	17	37	57	77	97	117	137	157	177	197	217	237	257	Ollin
Etznab	18	38	58	78	98	118	138	158	178	198	218	238	258	Tecpatl
Cauac	19	39	59	79	99	119	139	159	179	199	219	239	259	Quiahuitl
Ahau	20	40	60	80	100	120	140	160	180	200	220	240	260	Xochitl

First World of the serpent   Second World of the dog   Third World of the eagle   Fourth World of the sun

Figure 6.6. Tzolkin chart with the overlay of the filtration pattern of four Worlds. The four Worlds are named for their ending day signs.

Tzolkin Chart 7
with Five Worlds

Mayan Day Signs   *Aztec Day Signs*

Sign		1	2	3	4	5	6	7	8	9	10	11	12	13		Sign
Alligator		1	21	41	61	81	101	121	141	161	181	201	221	241		Alligator
Wind		2	22	42	62	82	102	122	142	162	182	202	222	242		Wind
Night		3	23	43	63	83	103	123	143	163	183	203	223	243		House
Seed		4	24	44	64	84	104	124	144	164	184	204	224	244		Lizard
Serpent		5	25	45	65	85	105	125	145	165	185	205	225	245		Serpent
Death		6	26	46	66	86	106	126	146	166	186	206	226	246		Death
Deer		7	27	47	67	87	107	127	147	167	187	207	227	247		Deer
Rabbit		8	28	48	68	88	108	128	148	168	188	208	228	248		Rabbit
Water		9	29	49	69	89	109	129	149	169	189	209	229	249		Water
Dog		10	30	50	70	90	110	130	150	170	190	210	230	250		Dog
Monkey		11	31	51	71	91	111	131	151	171	191	211	231	251		Monkey
The Road		12	32	52	72	92	112	132	152	172	192	212	232	252		Grass
Reed		13	33	53	73	93	113	133	153	173	193	213	233	253		Reed
Jaguar		14	34	54	74	94	114	134	154	174	194	214	234	254		Ocelot
Eagle		15	35	55	75	95	115	135	155	175	195	215	235	255		Eagle
Vulture/Owl		16	36	56	76	96	116	136	156	176	196	216	236	256		Vulture
Earth		17	37	57	77	97	117	137	157	177	197	217	237	257		Movement
Flint		18	38	58	78	98	118	138	158	178	198	218	238	258		Knife
Rainstorm		19	39	59	79	99	119	139	159	179	199	219	239	259		Rain
Light/Lord		20	40	60	80	100	120	140	160	180	200	220	240	260		Flower

Figure 6.7. Tzolkin chart with an overlay of the filtration pattern of five
Worlds. These five Worlds were ruled by various mythological themes:
the First World was ruled by creation, the Second World by water,
the Third World by burning water, the Fourth World by blossoming war,
and the Fifth World by the unification of opposites.

	National Underworld (3115 BCE–2011)	Planetary Underworld (1755–2011)	Galactic Underworld (1999–2011)	Universal Underworld (2011)
First World	3115–1834 BCE	1755–1819 CE	Jan. 5, 1999–March 20, 2002	Feb. 11, 2011–April 16, 2011
Second World	1834–552 BCE	1819–1883	March 20, 2002–June 2, 2005	April 16, 2011–June 20, 2011
Third World	552 BCE–730 CE	1883–1947	June 2, 2005–Aug. 15, 2008	June 20, 2011–Aug. 24, 2011
Fourth World	730–2011	1947–2011	Aug. 15, 2008–Oct. 28, 2011	Aug. 24, 2011–Oct. 28, 2011

Figure 6.8. The time spans of the Four Worlds in the National, Planetary, Galactic, and Universal Underworlds

Worlds were believed to have preceded what they saw as the present Fifth World. In their view Tonatiuh and the tzolkin symbol 4 Ollin (movement), which is identical with 4 Caban (earthquake) of the Maya, ruled the present World in the center. Around this center are the four different tzolkin symbols of 4 Ocelotl (4 Jaguar/Ix), 4 Ehecatl (4 Wind/Ik), 4 Quiauitl (4 Rainstorm/Cauac), and 4 Atl (4 Water/Muluc). Each of these tzolkin symbols corresponds to what was believed to have been a previous World, which had been destroyed in a catastrophe of some kind. The First World began with the tzolkin energy 4 Jaguar and ended when jaguars ate its giant inhabitants. The Second World of 4 Wind was destroyed by wind and its inhabitants were turned into monkeys. The Third World, ruled by 4 Rainstorm, was destroyed by fire and its inhabitants turned into turkeys, while the Fourth World of 4 Water was destroyed by floods and its people turned into fish. After this the present Fifth World of 4 Ollin was created, which is prophesied to be destroyed by famine and earthquakes, as its tzolkin sign also seems to imply.

What are we to make of this? To begin with, the Mexicas sought to use the tzolkin as a tool for prophecy, but not in a very successful way, since the prophetic information in the Calendar Stone has been very much scrambled. The Mexicas were somewhat in the same situation as

Figure 6.9. The Aztec Calendar Stone, with the sun god Tonatiuh at the center surrounded by the day signs of the preceding four Worlds

we are today in trying to reconstruct the true meaning of the Mayan calendar. Thus they had inherited most of their calendrical knowledge from the Toltecs, whose books they had burned.

At a deeper level, the reason that the information from the Mexicas cannot be taken at face value is that the consciousness of time in the world had been altered in 730 C.E., with the beginning of the real Fourth World of the Great Cycle—that is to say, its last sixty-five katuns. In this Fourth World not even the Maya could see the spiritual energies of time clearly, so calendars succumbed to materialist illusions. From this point onward calendars in Mesoamerica deteriorated. The Fourth World of the Great Cycle meant a first pulse toward the blindfolded consciousness of the Planetary Underworld shown in figure 5.13, page 114. As a result, the

Long Count was abandoned and the tun-based system was gradually replaced by physically based fifty-two-year calendar rounds, which unfortunately epitomizes the Mesoamerican calendar system. Thus we must be cautious about the scrambling effect on the calendrical information to which the Fourth World of the Great Cycle might have given rise. Although the Mexica, living far into this Fourth World, had glimpses of prophetic understanding, they did not develop a calendar that we can today recognize as a reliable description of the cosmic plan. A typical example of how the information from the Mexica Calendar Stone has been scrambled is that between some of the tzolkin combinations that mark the Five Worlds there are fifty-two units, while between others there are sixty-five. The Calendar Stone cosmology actually seems to reflect a mixture of Four Worlds of sixty-five units each and Five Worlds with fifty-two units each.

Another example of how the information from the Mexicas has been scrambled is how these previous Worlds were dated. The First World supposedly lasted from 955 to 279 B.C.E. (in the Gregorian calendar), the Second from 279 to 85 C.E., the Third from 85 to 397 C.E., and the Fourth from 397 to 1073 C.E., at which time the present World of 4 Ollin supposedly began. This means that they thought the First World lasted for thirteen fifty-two-year cycles, the Second seven fifty-two-year cycles, the Third six fifty-two-year cycles, and the Fourth again thirteen fifty-two-year cycles. The subdivisions into Worlds were not made evenly.

This scheme is clearly not one of true prophecy because it was based on the fifty-two-year cycle, which played a prominent role only among the later postclassical Maya and the Mexicas. While this fifty-two-year cycle is partly based on the tzolkin, it is fundamentally based on the physical 365-day *haab* year and so lacks prophetic relevance. The Mexicas were thus in an intermediate position between the classical Maya and the modern world in terms of prophecy. They retained some prophetic knowledge from the tzolkin and the energies of their calendrical deities. But the beginning of the dominance of the physical cycles, which in today's world has become almost total, led them astray, and so they were not able to precisely describe the cosmic time plan.

There is an interesting piece of information, however, that can be retrieved from the Aztec scenario: the Second and Third Worlds are based on the equation $7 + 6 = 13$. Thus the Aztecs are here conveying to us some idea of the fact that Thirteen Heavens could be divided into seven Days and six Nights. It is significant that the Second World of seven cycles was ruled by Quetzalcoatl, the lord of light, and the Third World by a deity we now know to have been a representative of darkness, Tlaloc. This is an indication that the Mexicas also knew that the Thirteen Heavens could be divided into seven cycles of light and six of darkness. The method of dividing these pulses of light may not have been correct, nor was their dating, but nonetheless this is evidence of another ancient people that was intuitively aware of the basic rhythm of divine creation.

7

The Galactic Underworld

THE EMERGING GALACTIC UNDERWORLD

This cycle [of thirteen tuns] will begin in 1999.
In the same way as the written language was used only
by a small part of the human population 13×20^2 tun
ago, and is only today in use throughout the world, we
may assume that the possibility that will emerge in
1999 will make a start that is relatively little noticed
and then rapidly spread across the planet to prepare
for the end of the cycle in 2011.

CARL JOHAN CALLEMAN, *MAYA-HYPOTESEN*

As of January 5, 1999, humanity has entered the Galactic Underworld (fig. 7.1). This is the eighth of the Nine Underworlds, that frame of consciousness is developed by a sequence of 13 tuns = 4,680 days. As we saw earlier, this is an Underworld that, because of its dominating yin/yang polarity, will tend to strengthen the East and develop the aspects of the human

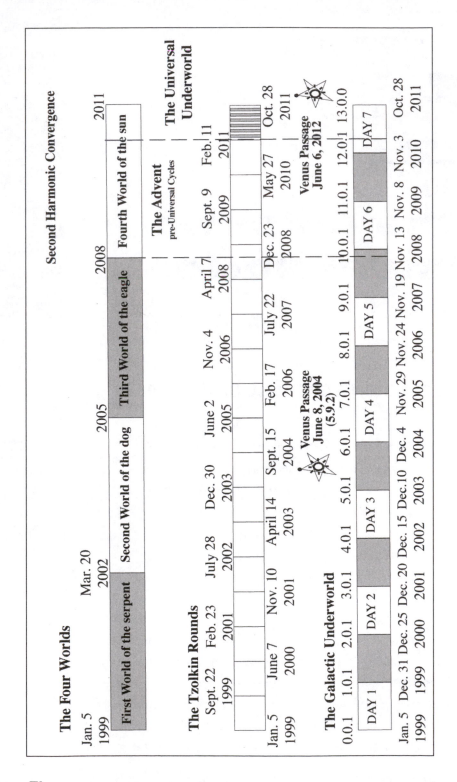

Figure 7.1. The Galactic Underworld with its Days and Nights (bottom row) running parallel to its tzolkin rounds (middle row) and Four Worlds (upper row)

psyche that are linked to the right-brain hemisphere. What kind of phenomena will then be developed in this cycle? Because nothing is totally new under the sun, and most phenomena are prepared for by earlier cycles, a good way to find out is to examine the phenomena that emerged around the time of the beginning of the Fourth World of the Planetary Underworld, on October 4, 1947 (see fig. 6.8, page 133). The Fourth World has prepared the ground for the Galactic Underworld, and so at the beginning of the Fourth World we can discover the embryonic forms of the phenomena that will come to dominate the Galactic Underworld.

An aspect of the Galactic Underworld that is easy to trace back to the beginning of the Fourth World is the current information technology (IT) civilization. The first computers were invented in 1946–1948, yet only with the First Day of the Galactic Underworld (January 5, 1999–December 30, 1999) did the idea spread widely that what was emerging was a "new economy" and a new type of civilization. It was typically declared that the changes the IT revolution brought about were on a scale commensurate with those of the Industrial Revolution in the mid-eighteenth century. From the perspective of the Mayan calendar, this parallel between the two revolutions seems very appropriate, since they were both seeded by Xiuhtecuhtli as the first Days of two different Underworlds began. Information technology is as much a part of the cosmic plan as was the invention of writing or the telegraph.

The IT civilization may be one of the aspects of the Galactic Underworld that is still most visible in the West. It builds on the type of science and technology that was developed especially by that hemisphere in the National and Planetary Underworlds. Yet right-brain hemisphere characteristics, which are typical of the Galactic Underworld, are also evident in IT phenomena, which possess a strong element of magic. These phenomena were developed through a creative process that is more intuitive than the traditional scientific method of the National Underworld, and the appeal of the IT world to the young is obvious to everyone. Pulse by pulse, major new innovations in this area and the ups of the "new economy" will be implemented during periods that are Days. At this point (2003) we are, however, still at a very early point in the development of this Underworld (corresponding, for instance, to the optical telegraph in figure 5.12, page 109).

Another technology that will be carried by the Galactic Underworld is that of genetics, and we may track its embryonic beginning back to the first DNA transfection experiments in the late 1940s. The completion of the Human Genome Project is an early expression of the current Underworld, and certainly the uses of genetic manipulation will remain a matter of heated debate.

In the media, the focus has mostly been on technologically and economically important aspects of the emerging Galactic Underworld. It would be a grave mistake, however, to conclude from this that these are also the most important. Even if some technologies will continue to advance during the Days of this Underworld, the tremendous attractive power these have on many can be expected to recede as we move farther into the Underworld.

The Galactic Underworld is more about telepathy and intuition than technology, and although many will resist the change, the emphasis of the future will be less on things and gadgets, as was the case in the Planetary. Because of the blindfolded consciousness that ruled the latter, all nature was assumed to exist only for human beings to exploit based on the economic calculations of the left-brain hemisphere. Although the intuition of the right-brain hemisphere in the Galactic Underworld may indeed lead to technological advances, the overall emphasis of human endeavor is now about to shift. Despite humankind's immense technological advances up until now, the advances it has made in attaining an inner state of happiness have been very small, and this will become increasingly evident. The main purpose of the current Underworld is to remedy this imbalance, this state of Koyaanisqatsi. The frame of consciousness that is now emerging will simply not allow us to deny the existence of a living cosmos. If we do, there will be a backlash of some kind. This is the reason for the urgency of delivering the message of the Mayan calendar.

Because of the nature of this Underworld's yin/yang polarity, its

Days will bring forth intuitive, rather than analytical, expressions of the human mind. Such expressions will rarely be predictable by a narrowly rational mind and may often seem to appear out of the blue. Nonetheless, we will see a return to an enchanted cosmos, and we will, through our increasingly more open senses, become sensitive to its true spiritual energy. Because of their different locations in relation to the World Tree, however, different parts of the world are likely to be affected differently by the Galactic Underworld. Thus we may expect, at least initially, that in the Western Hemisphere the Galactic Underworld will generate a fascination with the new technology and the new economy, while the focus in the East—India, China, and Russia (the leading edge will gradually move north)—will be directed more toward spiritual enlightenment. As the Galactic Underworld progresses, the effects of the energy shifts between different tuns—Days and Nights in the creation sense—will be increasingly more marked.

The points made so far about the emerging Galactic Underworld are based on the new yin/yang polarity by which it is dominated (see fig. 5.13). Another of its aspects is the widening of the human frame of consciousness. Nations as political and economic units are disappearing. Day by day the frame of human consciousness will expand in the Galactic Underworld, and limiting structures with an origin in lower Underworlds will be transcended. In a world without boundaries, calls for patriotism will become increasingly hollow. This aspect of the "galactification" process is today moving most quickly in Europe, where individual nations in the traditional sense hardly exist anymore. The European Union has absorbed the nation-states to a much higher degree than most of their inhabitants would probably like to think. There nationhood is primarily expressed through different soccer teams, special holidays, and so forth, but has little or no meaning in the world of politics and economics. Of course, as national boundaries are dissolved, armed conflicts and wars will be fought much less between different nations and increasingly between groups that are in resonance with different energies, regardless of what country they may happen to live in (e.g., stateless terrorism carried out against the West).

Thus everything is happening on an increasingly larger scale. No longer are the nations the most significant players in the arena of global politics. A series of protests, beginning in Seattle in 1999, in the first Day of the Galactic Underworld, have highlighted that the World Trade Organization, International Monetary Fund, World Bank, Group of Eight, European Union, United States (which, in the global political arena, is not a nation in the traditional sense), NATO, and various multinational corporations have replaced the nations as the most powerful players in the global political arena. Although the new countermovement has mostly been referred to as antiglobalist, in light of the emerging Galactic Underworld it is probably better described as a movement against the dominance of materialist values on our planet.

If the Eastern Hemisphere is strengthened in the Days of the current Underworld, this will also mean that the qualities of the right-brain hemisphere will be favored, and so the current global dominance of the left-brain hemisphere and the Western Hemisphere will come to an end. This change will affect everyone, regardless of where he or she happens to live. We are all one and are all products of the same divine process of creation, so it is no wonder that in recent times many, regardless of location, have been working to balance the two hemispheres of their brains by developing their intuition. As a consequence of the receding dominance of the left-brain hemisphere, the controlling mentalities it carries will also come to an end. As the brain hemispheres become equal, so will human individuals, and as a consequence no soul will be able to control another soul. In general, an irreversible evolution toward wholeness will take place as the Galactic Underworld progresses, and in the process all hierarchies based on dominance—political, religious, or otherwise—will, in one way or another, break down. Needless to say, this process will hardly be smooth or without resistance on the part of the established world order.

Because a dualist frame of consciousness rules the Galactic Underworld, it does hold a potential for violent conflicts and wars, and we may well have entered a negative spiral of terrorism and terror against terrorism in which the number of survivors is anybody's guess. The

Galactic Underworld will develop through a wave movement whose Days will essentially strengthen the right brain and the Eastern Hemisphere, while the left brain and the Western Hemisphere will seem to regain lost ground during the Nights. The current materialist civilization will crumble, so clinging to the old ways will not give security. Much turbulence is to be expected, for the simple reason that a significant energy shift will take place at every tun and will seem to pull the world in a new direction. Moreover, the Galactic Underworld develops at a high tun-based frequency during which the control of the world by the hierarchies generated by lower Underworlds will come to an end in a very short period, about ten years. The reason that equality will be the end result is that as this Underworld draws to a close, the light will be evenly distributed across the planet.

GUIDANCE FROM HISTORICAL ANALOGIES

Of course, the new yin/yang polarity of the Galactic Underworld will not necessarily manifest in acts of terrorism or wars. But we have reason to expect a general potential for conflicts between people that are primarily in resonance with either of the two hemispheres, regardless of where on the planet they happen to live. Another way of looking at this is that the emerging yin/yang polarity may be expressed as a conflict in each and every individual between the aspects of our existence that are mediated through the left- and the right-brain hemispheres. Needless to say, there is potential for an apocalyptic scenario, and quite possibly World War III has already began; it just does not look like the preceding ones. Yet, as I will discuss later, there is also the possibility of unifying the opposing polarities, and it is from such intent that the hope for humanity comes. A great unknown is to what extent and how soon people will become aware of the process of divine creation and its time plan provided by the Mayan calendar. This awareness could by itself have a profound influence on the course of events.

The Mayan calendar allows us to make predictions for the future based on historical analogies. The reason such predictions are possible is that the thirteen deities ruling the progression through the Thirteen Heavens of an Underworld are always the same. This is exactly the principle upon which the ancient Mayan art of prophecy, described in the Books of Chilam Balam, was based. We may thus understand much of the current and future course of events through parallels between the periods that the thirteen deities rule in different lower Underworlds (fig. 7.2).

Learning from historical analogies, however, requires that the beginning and end dates of each Underworld are correctly calibrated. In our current Galactic Underworld, where shifts between Days and Nights take place as frequently as every tun, this becomes even more important than in the National Underworld, for example, where energy shifts take place every baktun and a minor error in the end date plays less of a role. Nonetheless, if the end date of the Galactic Underworld is incorrect, the possibility of predicting changes in consciousness from the Mayan calendar would be completely lost and its practical use rendered meaningless. (The same is true if you use a tzolkin count that is not a true reflection of divine creation.) A difference of a year when it comes to the end date means all or nothing for those seeking to align themselves with the wave movement of cosmic evolution. Other than myself, however, probably only Solara and Ken Carey, both approaching the matter from an intuitive rather than a scholarly point of view, have clearly specified the time that unity will be attained at the end of 2011.

It should be noted that the completion date of creation given here—October 28, 2011—deviates from the ending of the Mayan Long Count of December 21, 2012. The latter date is the one that we still most commonly hear, and it is indeed correct as far as archaeology goes. To understand the divergence between these dates, note that the ending date of the Long Count depends on its date of beginning, which was set by the Maya. The ancient Mayan inscriptions do not tell of what will happen at the end of the Long Count, but about what happened at its beginning, when the First Father raised the World Tree. This day was August 11, 3114 B.C.E., the exact date of which, however, seems to have been based on an old tradition in the location of Izapa,

Day/Night Duration	Ruling Aztec Deity—Growth Stage Phenomenon or Prophecy
Day 1 Jan. 5, 1999– Dec. 30, 1999	**Xiuhtecuhtli—Sowing** IT revolution; living on light; wars of the West with Iraq and Serbia; first protests (in Seattle) against global materialist power
Night 1 Dec. 31, 1999– Dec. 24, 2000	**Tlaltecuhtli** Backlash of IT revolution; resting period following the initial day of the Galactic Underworld
Day 2 Dec. 25, 2000 Dec. 19, 2001	**Chalchiuhtlicue—Germination** Continuing protests against global materialist power; terrorist attacks against the World Trade Center and Pentagon; Western war in Afghanistan
Night 2 Dec. 20, 2001– Dec. 14, 2002	**Tonatiuh** Deep economic recession; reestablishment of Western dominance; reaction against most expressions of the high frequency of the Galactic Underworld; cultural search for answers in ancient cultures
Day 3 Dec. 15, 2002– Dec. 9, 2003	**Tlacolteotl—Sprouting** Western war in Iraq; strengthened movement against global power of materialism; establishment of new expression of spirituality: intuition-based living, dowsing, revival of Mayan calendar, spiritual healing
Night 3 Dec. 10, 2003– Dec. 3, 2004	**Mictlantecuhtli** First Venus transit (June 8, 2004), crucial for preparation of the unifying synthesis of Day 4
Day 4 Dec. 4, 2004– Nov. 28, 2005	**Cinteotl—Proliferation** Formulation of core unifying synthesis between East and West and between spirituality and rationality (see fig. 4.6, page 81)
Night 4 Nov. 29, 2005– Nov. 23, 2006	**Tlaloc** Dissemination of unifying synthesis
Day 5 Nov. 24, 2006– Nov. 18, 2007	**Quetzalcoatl—Budding** Breakthrough of new, advanced form of unifying synthesis (manifestation of galactic Christ consciousness)
Night 5 Nov. 19, 2007– Nov. 12, 2008	**Tezcatlipoca** Deep crisis for global materialistic culture; destructive reaction; "Armageddon"; European Union and Germany in center of drama
Day 6 Nov. 13, 2008– Nov. 7, 2009	**Yohualticitl—Flowering** Renaissance of advanced unifying synthesis; tense coexistence of East and West and between new spirituality and the remnants of the global materialistic power
Night 6 Nov. 8, 2009– Nov. 2, 2010	**Tlahuizcalpantecuhtli** Resting period in preparation for attaining balance between East and West; Second Harmonic Convergence (May 27, 2010), the first experience of a galactic pulse of cosmic consciousness
Day 7 Nov. 3, 2010– Oct. 28, 2011	**Ometeotl/Omecinatl—Fruition** Balance attained between East and West; end of control by one soul of another soul; beginning of the Universal Underworld (Feb. 11, 2011) and the development of a nondual cosmic consciousness

Figure 7.2. Prophetic progression through the Galactic Underworld

where the Long Count was invented. In accordance with this tradition, time had begun on the day of the year corresponding to August 11, when the sun was at its zenith in this location.

In this cradle of the Long Count, this day in the solar year had already become a holy day ("when time began") that could not be questioned (think how difficult it would be to change the date of Christmas, even if it were proved beyond a doubt that Jesus had not been born then). If the exact beginning date of the Long Count was based on a very local tradition in Izapa, it has no relevance as we apply the Mayan calendar on a global scale. Through fairly extensive research, which is outside the scope of this book, I have come to the conclusion that the correct date for the completion of creation is October 28, 2011, a day that has the energy 13 Ahau. This 13 Ahau, which is the last day of a tzolkin round, is when the light will pass through all Underworlds without any obscuring filter of darkness blocking the contact between humankind and the Divine.

Based on this completion date for the process of divine creation, it is possible to learn something from historical analogies about the wave movement of the Galactic Underworld through its thirteen ruling deities.

The evolution of the new consciousness in this Underworld will, as in all others, be brought about by a wave movement alternating between Days (external change, creativity) and Nights (internal change, rest). During the Days of this Underworld intuitive ways of thinking and acting will be strengthened, or made visible, while during the Nights the left-brain thinking that we already know from the National and Planetary Underworlds will resurface. By definition this wave movement also means that evolution will not be linear. It is never possible to understand the course of events simply by projecting trends linearly into the future.

With every Day the new galactic frame of consciousness will be strengthened at the expense of the more limited planetary and national frames. The development of the Galactic Underworld not only entails the potential for conflicts between East and West, but also between the phenomena that were developed by the lower Underworlds and those emerging with the Galactic Underworld. The phenomena developed by the Galactic Underworld will advance according to the pulsewise Day-by-Day pattern, of which a few examples from lower Underworlds have been given in this book. The Nights in between serve essentially for the integration of the pulses of the previous Days and as preparation for those to come, and hence may be the best for contemplation. As we progress deeper into the Galactic Underworld, these energy shifts between Days and Nights will become increasingly more marked, and will make more distinct the guidance from historical analogies using the matrix based on my research (which I've here termed, for lack of an easier identifier, the Calleman Matrix).

If we consider the turbulence and the amount of violence that arose as the Planetary Underworld began—the Seven Years' War, the American and French Revolutions, and the Napoleonic Wars—it is obvious that the beginning of a new Underworld is hardly smooth. What may we then expect from the further development of the now emerging Galactic Underworld based on our knowledge of the energies of the Thirteen Heavens? In *Solving the Greatest Mystery of Our Time: The Mayan Calendar,* published in early 2001 but written before the onset of the Galactic Underworld, I made a few predictions based on comparisons between Underworlds.

First, I stated that because of the increased frequency of change of this creation cycle (shifts between light and darkness taking place every 360-day tun rather than every 19.7-year katun), time will be perceived as accelerating. Events would thus unfold at a faster pace than has ever been experienced before. Many people would probably agree to the existence of such a speeding up of time.

Second, I predicted that—at least in the Northern Hemisphere—the world would come to be divided into three main sections: (1) the Western Hemisphere, including the British Isles; (2) central Europe, under the trunk of the World Tree; and (3) the Eastern Hemisphere, including Russia. This new division manifested in the very first weeks of the first Day of the Galactic Underworld, as the United Kingdom and the United States jointly attacked Iraq. We saw this alliance also in the war against the Taliban. The notion of West is thus now becoming more specific and more clearly related to the World Tree. To make this difference even more pronounced, central Europe has very rapidly been moving ahead toward its unification in the European Union, and at least at present, the odds seem to be against the United Kingdom changing currency and thus fully becoming part of that union.

The third prediction I made is that the current Underworld would be dualist, thus generating conflicts between the East and the West. This is in contrast to the period 1992–1999, which was by far the most peaceful in the history of humankind. In this period there were no wars between different nations (only civil wars). The reason that this period was so relatively peaceful was that in 1992 we arrived at the seventh Day of the Planetary Underworld (which was a unitary Underworld), while the dualist Galactic Underworld had not yet started. The prediction of increased conflicts between East and West after 1999 has since been verified not only in the NATO war with Yugoslavia in its first Day, but also in the attack by Islamic terrorists on the World Trade Center and Pentagon in the United States, the Western war in Afghanistan, and an intensification of the Israeli-Palestinian conflict in the second Day. A further prediction that has been verified is that these wars have manifested in periods that are Days rather than Nights. The Western war in Iraq in the Third Day (see appendix D) has further confirmed this. As we understand from the Mayan calendar, this is because the Days are ruled by the new yin/yang polarity, which holds a potential for conflict.

The Galactic Underworld has thus begun with an intensification of the conflicts between the East and the West, including between the Western Chaldean-Jewish-Christian tradition of monotheism and its Eastern Muslim counterpart. Yet, as we will discuss later, it may also lead toward an integration of some of the ancient spiritual traditions of the two hemispheres. Europe is then likely, from Day Four (the Seventh Heaven) onward, to take on a special role in unifying this polarity. Integrated expressions of the left- and right-brain hemispheres will emerge in Europe and spread via the World Tree. The effects of the

World Tree, essentially invisible in the Planetary Underworld, will again be evident in this Underworld.

Hence *light* or *Day* does not necessarily mean "good" and *darkness* or *Night* does not need to mean "bad." It is more complex than that. In lower Underworlds there are many examples of how wars or violent conflicts have erupted during Days, periods of light. Thus in the dualist National Underworld, wars were notorious during all the Days. In the Planetary Underworld, when at least an ideal of peace started to emerge, both the Napoleonic Wars and World War I occurred in periods that were Days, Day Two (1794–1814) and Day Five (1913–1932), respectively. How could this be? Part of the answer lies in the fact that the change in energy associated with a change between Heavens always holds the potential for shifting the earthly balance of power. In addition, historical events must always be understood in terms of the purpose of the particular Underworld from which they result. The divine process of creation operates like the Hindu god Shiva, the Creator and the Destroyer, who destroys in order to create the higher purpose, regardless of what human beings may desire. Ultimately, the course of events is subordinated to the intended end result of the cosmic plan, which is to be manifested in 2011.

Part of the purpose of the Planetary Underworld was, for instance, to develop democracy, a first step toward the full equality of all human beings that will be attained only as the Universal Underworld draws to a close. Equality had been completely negated in the National Underworld, where one was born into an essentially static class society, a separation of people caused by the Fall that had dominated human societies ever since the beginning of the dualist National Underworld. From the perspective of the Planetary Underworld, it is easy to see that both the Napoleonic Wars and World War I, despite the tragedies and the human lives lost, created the space for significant steps toward the emergence of the new democratic system of rule, a first step away from the fixed class society of the National Underworld. The Napoleonic Wars in Day Two, with their origin in the French Revolution, shook the established order of the royal houses of Europe in such a way that that order could never again be taken for granted. Napoleon was seen as a terrorist or worse, especially by the royal houses, who at the time had the most to lose (epithets for Napoleon in a royalist newspaper at the time included "the Man Eater," "the Werewolf," and "the Beast").

World War I in Day Five, in which tens of millions lost their lives, meant a more definitive collapse of the established order. Several of the autocratic empires that had ruled the world—the Russian, Chinese, Ottoman, Austrian, and German—came to an end, and in the aftermath of this war a wave of democracies was instituted in Europe. It is highly questionable whether this would have happened had the war not ended the rule of these empires. This does not mean that Napoleon or any of the generals of World War I could be described as men of democracy or equality, only that the cosmic plan sometimes works its way to its purpose through people who have other goals.

And so, based on the parallels between the ruling deities, the changes propelled by Days Two and Five, for instance, could also generate violence in the Galactic Underworld. In the case of Day Two (December 25, 2000–December 19, 2001), we have already seen examples of this. Because this time period was dominated by the same deity (Chalchiuhtlicue) of the Thirteen Heavens as the period (1794–1814) in the Planetary Underworld, it is natural to discuss parallels between Napoleon Bonaparte and Osama bin Laden. First of all, the scale of the ambitions of the two individuals does not seem to have been much different. Napoleon's ambition was for his imperial brand of the French Revolution to conquer the world, and supposedly bin Laden wanted Islam to do the same. Millions died in the Napoleonic Wars, and several thousand innocent civilians were murdered in bin Laden's terrorist attacks; both men, if it served their ideals, seem to have lacked concern for the loss of human lives, both in their own camps and among the enemy.

What may be more important than these similarities is that the two men were both surfing on the early (Day Two) waves of the respective Underworlds by which they were dominated. Thus they were carriers of very distorted expressions of the phenomena that these Underworlds are ultimately meant to develop. While Napoleon was a child of the French Revolution, he himself betrayed its egalitarian ideals by becoming an emperor. Put in other terms, Napoleon had very little to do with the kind of democracy for the later development of which he paved the way. Similarly, bin Laden may be part of a process serving to strengthen the East and the right-brain hemisphere, but he and the Taliban clearly did not embrace the ideals of global unity or equality between genders whose development is the ultimate purpose of the Galactic Underworld. Just as Napoleon met his Waterloo (1815) as the Second Night of the Planetary Underworld began, bin Laden disappeared from political sight, and may well have been killed, at about the time that the Second Night of the Galactic Underworld began on December 20, 2001. Nonetheless, both shook the established order in the Second Day of their respective Underworlds.

In the Planetary Underworld, Night Two (1814–1834) was the historical period that has been called "the Reaction." Following the fall of Napoleon, the great powers of Europe gathered at the Vienna Conference in 1815. There the royal houses that had ruled Europe prior to all the preceding turbulence were reinstalled. This was a reaction to all the preceding changes that the Industrial Revolution, as well as the American and French Revolutions, had brought about in what then seemed like a very short period of time. Culturally, Night Two of this cycle was characterized by Romanticism, a longing to return to a mythical past, as a reaction to the novelties that the Planetary Underworld had brought. This was, among other things, the period when the sagas of preindustrial Europe were again beginning to be read.

In Night Two of the Galactic Underworld, parallel phenomena also appeared. The old order of the economic, military, and media dominance of the West was firmly reestablished, and people were nostalgic for the time before all the changes that the IT revolution, terrorism, and so forth had produced at an alarming rate. *The Lord of the Rings: The Fellowship of the Ring,* for instance, had its world premiere on the first day of the new Night, and *Harry Potter and the Sorcerer's Stone* a week before, both expressions of the Romanticism of the Galactic

Underworld—or, to be more precise, the Romanticism of Tonatiuh (see fig. 2.2, page 19). There was also a search for solutions in the ancient cultures of humanity, together with an economic recession and a growing realization that sustained economic growth was a thing of the past.

The Third Day in all Underworlds is when the phenomena that they develop are truly anchored in physical reality. In the Galactic Underworld this energy covered the period December 15, 2002, to December 10, 2003. Among the examples of evolutionary progressions given earlier, Moses' monotheism, alphabetic writing, and the electrical telegraph were expressions of the consciousness of Day Three. Day Three (the energy of Tlacolteotl) is thus the energy among the seven Days of an Underworld when lasting, and sometimes practically useful, solutions manifest. In fact, the examples of the manifestations of the Third Day given above are still part of today's world, whereas the expressions of earlier Days, such as cuneiform or the optical telegraph, have become obsolete. Such comparisons also tell us that we are still early in the Galactic Underworld and that the expressions developed before the Third Day of an Underworld are really very rudimentary compared to what is to come.

It seems likely that the right-brain hemisphere will be expressed more broadly, including in people in the Western Hemisphere. Although on one level it may appear that the most important event in the first half of the Third Day was that the West won another war, the lasting effect of this war may be the great opposition to it, even among countries that have traditionally been seen as allies of the West. Again, even if the actual detailed developments are unpredictable, we know from the Mayan calendar that in one way or another the mentality of the Eastern Hemisphere will assert itself during this period, because this is a new Day. Moreover, in the Third Day of the Galactic Underworld it will assert itself in such a way that the Western dominance of the world will begin to be seen as a thing of the past. People everywhere will come to realize that they are living in a new reality, a new Underworld. Many old values will collapse.

It would be going too far to discuss what we might expect from all the Thirteen Heavens of the Galactic Underworld. The reader has been provided with the tools to make his or her own studies using parallels between Underworlds as a basis. In figures 7.3–7.5 the Planetary and Galactic Underworlds have been matched energy-wise with the corresponding time periods. The examples I use from the evolution of the Planetary Underworld are some significant steps in the development of democracy taken during its Days. The reader may use the parallel row for the Galactic Underworld to fill in the steps in the evolution of contacts with extrasolar civilizations, the integration of the Eastern mind, or any other development of interest.

I will, however, discuss a few more examples of how the Mayan calendar may be relevant for religious prophecy. The use of historical analogies to follow the course of events in the current Galactic Underworld may also have some bearing regarding the prophecy of the return of Christ. Is there a reason to expect that Christ will return, as is prophesied in the Bible? From the perspective of the Mayan calendar, the teachings of Jesus and the rise of Christianity were primarily a function of the consciousness generated by the Fifth Day of the Great Cycle, a Day ruled by Quetzalcoatl, the lord of light. The change in consciousness brought about by the Heaven of Quetzalcoatl created receptiveness to the Christian message among the people. We would expect that in the corresponding time period, the Fifth Day, of the Galactic Underworld (approximately Gregorian year 2007), a new expression of Christ-consciousness will emerge. This might be called the return of Christ, although it seems wiser to place the emphasis on the energy change than on the appearance of a certain individual. In the Galactic Underworld, however, the Fifth Day will carry the yin/yang duality opposite to that of the National Underworld. Thus in the Galactic Underworld the loving impulse will not be generated by its dualist consciousness. Galactic Christ-consciousness will instead unify the effects of the two yin/yang polarities and serve to make us whole, individually and collectively. The origin of this impulse will most certainly be the East.

There are some interesting cases of how certain individuals in the

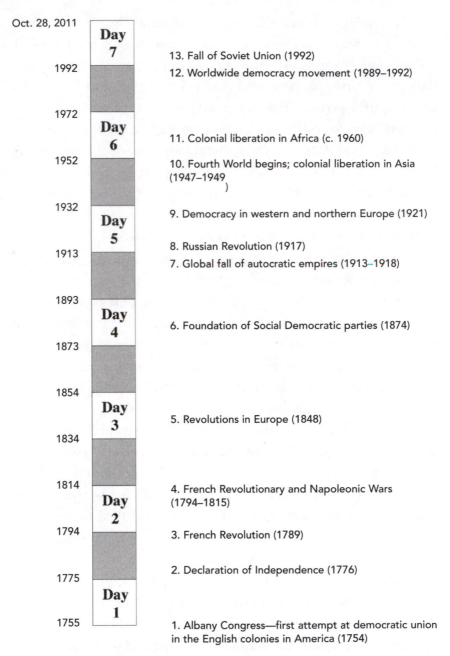

Oct. 28, 2011

Day 7

1992

13. Fall of Soviet Union (1992)

12. Worldwide democracy movement (1989–1992)

1972

Day 6

1952

11. Colonial liberation in Africa (c. 1960)

10. Fourth World begins; colonial liberation in Asia (1947–1949)

1932

Day 5

1913

9. Democracy in western and northern Europe (1921)

8. Russian Revolution (1917)

7. Global fall of autocratic empires (1913–1918)

1893

Day 4

1873

6. Foundation of Social Democratic parties (1874)

1854

Day 3

1834

5. Revolutions in Europe (1848)

1814

Day 2

1794

4. French Revolutionary and Napoleonic Wars (1794–1815)

3. French Revolution (1789)

1775

Day 1

1755

2. Declaration of Independence (1776)

1. Albany Congress—first attempt at democratic union in the English colonies in America (1754)

Figure 7.3. The pulsewise development of democracy in the Planetary Underworld. Although many of the Days of the Planetary Underworld seem to have manifested in unpredictable and disparate events, there is still a clear movement whereby, pulse by pulse and Day by Day, the previous rule of monarchs and nobility is replaced by democracy, republics, and national sovereignty of former colonies.

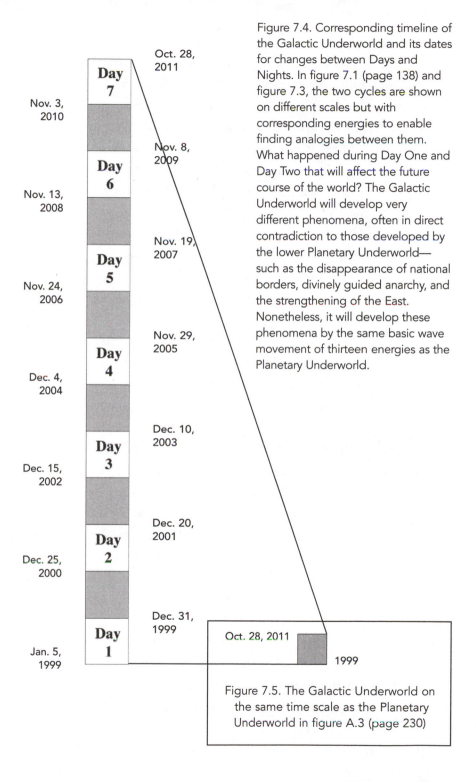

Figure 7.4. Corresponding timeline of the Galactic Underworld and its dates for changes between Days and Nights. In figure 7.1 (page 138) and figure 7.3, the two cycles are shown on different scales but with corresponding energies to enable finding analogies between them. What happened during Day One and Day Two that will affect the future course of the world? The Galactic Underworld will develop very different phenomena, often in direct contradiction to those developed by the lower Planetary Underworld—such as the disappearance of national borders, divinely guided anarchy, and the strengthening of the East. Nonetheless, it will develop these phenomena by the same basic wave movement of thirteen energies as the Planetary Underworld.

Figure 7.5. The Galactic Underworld on the same time scale as the Planetary Underworld in figure A.3 (page 230)

blindfolded Planetary Underworld were sensitive to the energies of the cosmic plan and were able on an intuitive basis to make correct historical analogies. One such example is the Jehovah's Witnesses, and another is Winston Churchill. The Jehovah's Witnesses had predicted that Jesus would return to earth in the year 1914 C.E. Instead, as we know, the world saw World War I followed by the Russian Revolution, and the group supposedly lost a high number of adherents because of the apparent failure of its prophecy. But from the perspective of the Mayan calendar, the Fifth Day (that of Christ-consciousness) of the Planetary Underworld actually began in 1913, and in this sense their prophecy was essentially correct—but in the Planetary Underworld people were blindfolded, and the consciousness of Quetzalcoatl manifested in much more materialist ways, such as in significant advances for democracy, science, and mass communications.

The other example is Winston Churchill, who talked about "Hitler and his Huns" in his radio speeches during World War II. The interesting parallel here is that when the Fifth Night (ruled by Tezcatlipoca, lord of darkness) of the National Underworld (434–829 C.E.) began, the Huns invaded central Europe. Later, when the Fifth Night (1932–1952 C.E.) of the Planetary Underworld began, Hitler became a dictator in the same region. In the National Underworld the rule of the lord of darkness (Night Five) generated the Dark Ages, whereas in the Planetary Underworld it generated the Great Depression, Nazism, Stalinism, the Holocaust, World War II, the Korean War, and the atomic and hydrogen bombs. Although the phenomena the two Underworlds carried had different characters, it seems clear that the period 1932–1952 was by far the darkest in modern history (corresponding to the Dark Ages of the National Underworld). We thus have reason to expect that Night Five in the Galactic Underworld, essentially the Gregorian year 2008, will turn out to be a very difficult time.

Although the destruction taking place during the Fifth Night serves to create the space for the new flower to blossom at the beginning of the Sixth Day, the rule of Tezcatlipoca in the Galactic Underworld is a period for which it would seem wise to be prepared. Just looking at the

Galactic Underworld from the perspective of economics, we can expect that increasingly the Nights will be downturns, while the Days may be times when the "new economy" is slightly more prosperous. The beginning of the Fifth Night, as with the collapse of the Roman economy in the National Underworld and the Great Depression in the Planetary Underworld, will be a major downturn. This will mean not only a major downturn in the capitalist economy, as was the case already in the Second Night of the Galactic Underworld, but more like a total collapse of the international monetary system and the beginning of the end of quantitative abstraction of values mediated by the left brain (see appendix A).

The rule of Tezcatlipoca (the Fifth Night) always means a last destructive attempt of the forces seeking a return to the rule of the previously dominating Underworld. Thus in the National Underworld, it meant a temporary return to the tribelike world of the nomadic Huns and the marauding Germanic tribes, distortions of the social structures that had dominated the lower Regional Underworld. In the Planetary Underworld, it meant, in a new guise (Hitler), a return to autocratic rulers, who had based their power on supposedly "superior blood." In principle Hitler's ideals were not different from those of the royal dynasties that throughout the National Underworld had based their power on having "blue blood" in their veins. By analogy, we can expect that the rule of Tezcatlipoca in the Galactic Underworld will mean a last violent assault by the forces seeking a return to the materialist values of the Planetary Underworld. Also in this Underworld the struggle of Night Five will probably have central Europe as its focus.

The case of the Jehovah's Witnesses' prophecy illustrates one further point: Even if it had been possible for this group to intuitively access the energy of Quetzalcoatl, Christ-consciousness, this energy is expressed in different ways in different Underworlds. In the Planetary Underworld the Quetzalcoatl energy was expressed through a blindfolded frame. In the Galactic Underworld the Quetzalcoatl energy will very likely be expressed through the integration of the new pulse of light shining on the Eastern Hemisphere with that already shining on the West. Hence in the Galactic Underworld this energy of Christ consciousness may be expressed very differently from what would be expected from a prophecy, such as the Book of Revelation, developed through the eyes of the National Underworld.

After the beginning of the Fourth World of the Galactic Underworld on August 15, 2008—toward the end of Night Five—things will begin to move much more smoothly for those seeking the enlightened path. At that point some embryonic expressions of the Universal Underworld will have already begun to manifest and allow us to see the light at the end of the tunnel. Because it will be part of the Fourth World, Night Six will not be as destructive as Night Five.

GALACTIC SPIRITUALITY

In addition to the effects the Galactic Underworld will have on our own planet, we may also expect the widening of our frame of consciousness to have effects on the way we see its place among the civilizations of our galaxy. The first real wave of reports of UFOs came in 1947, at the beginning of the Fourth World of the Planetary Underworld. Although I do not believe such reports reflect real visits from other planets, they nonetheless reflected a shift of focus to the possibility that civilizations exist in other solar systems. The Days of the Galactic Underworld especially will provide further information on this point. In its First Day, in August 1999, the first conclusive evidence was presented for the existence of extrasolar planets, and in its Second Day, in 2001, it became possible to identify a chemical component on such a planet. The natural prediction is that in the Third Day of the cycle, roughly the year 2003, there will be evidence of oxygen—in other words, of life—on an extrasolar planet. From this point on, our galactic understanding will increase in a pulsewise Day-by-Day fashion. The widened frame of consciousness of the Galactic Underworld will guide our vision in the right direction for making contact with the rest of the galaxy.

What type of extrasolar planets would we have reason to suspect support life? About this the Mayan calendar may have something to say.

Assuming that all planets harboring life are the products of the same creation, going back to the Big Bang, we would expect that life in all solar systems is now entering the Galactic Underworld in synchrony and approaching enlightenment in unison. If in all extrasolar civilizations the evolution of consciousness is generated by the same tun-based divine process of creation, then we would expect planets that harbor beings with a humanlike consciousness to revolve around their stars in approximately 360 days. Secondly, many, but not all, sunlike stars have a full sunspot cycle of twenty to twenty-two years. If this periodicity is indeed a reflection of the fact that our own sun and many other stars are mediating cosmic information with a katun frequency, this would mean that we would be most likely to find an evolution of consciousness similar to that on our own planet around stars with approximately this same sunspot cycle. (Note that physical cycles are sometimes imperfect reflections of the underlying consciousness cycles. Our physical year, for instance, is a few days longer than the tun. The full sunspot cycle, which in the last century has decreased to about twenty years and is heading for a maximum in 2011, may similarly be an imperfect reflection of the katun frequency.)

In the context of extraterrestrial life, because in the Planetary Underworld people had a blindfolded consciousness, their view of spirituality became rather peculiar in some respects. This Underworld generated the idea that things needed to be outrageous to be regarded as spiritual, and so ETs, conspiracies of blood-drinking reptiloids, Lemurians, and so forth attracted attention, usually without any kind of empirical verification whatsoever. It is understandable that the blindfolded consciousness of this Underworld produced such ideas. Since the spiritual dimension of the real world was then not apparent, it would have to be assumed that it existed "somewhere else," in some hidden realm.

As a result, throughout the Planetary Underworld (from 1755 on), something of a fantasy world of esoterica has been developed by a series of prophets or would-be prophets. It has been typical of the esoteric tradition that, although it may have upheld the knowledge of the existence of a spiritual domain, ideas about the cosmic plan were not grounded in the evolution of events in the real world. In contrast, the spirituality that is emerging with the Galactic Underworld will be about a change in our perception that will allow us to see the spiritual as well as the material dimension of reality. Not only much of traditional materialist science and organized religiosity but also many of the esoteric traditions will come to an end now with the Galactic Underworld.

I would also like to say a word about possible earth changes in the Galactic Underworld. It might seem as though the Mayan calendar had little to say about geophysical processes, but if our sun plays a role as a mediator of information from the galactic center, we would have to expect that the sun is now also beginning to mediate the tun period of the Galactic Underworld, which currently influences our minds. Since the sunspot cycle by itself is clearly influencing our weather and climate, a similar cycle operating with a tun frequency may also start to have an influence on our climate. After all, if everything is related through an all-pervading invisible ether, it would be surprising if the current change in the universal field of consciousness did not also have some kind of physical effects.

At least some minor earth changes may thus result from the high-frequency change in consciousness that the cosmos is now undergoing. But the reversal of cause and effect to generate a doomsday view where such earth changes are presented as the very purpose of this particular phase of creation is untenable. Any possible physical effects should be seen only as by-products of changes in the consciousness field, and most climatic change will still probably be caused by carbon dioxide emissions rather than some purported insidious divine plan for the destruction of the earth through geophysical changes.

Many of the discussions about earth changes go back to the famous American psychic Edgar Cayce, who, when asked in a reading what was going to happen on earth around the years 2000–2001, delivered the answer, "When there is a shifting of the poles. Or a new cycle begins." From what we now know, it seems very likely that what he intuitively accessed was not a shift in the magnetic polarity of the earth but a shift in the yin/yang polarity of consciousness and the beginning of the Galactic Underworld ("new cycle"), as is evident from the Mayan calendar.

The esoteric traditions may be full of such correct intuitions that have been misconstrued because of the shortcomings of the thinking that ruled at their times of origin. Just as an example, the theosophist leader Helena P. Blavatsky's idea that humanity passes through seven root races in its evolution may be a combination of a correct intuition of the fact that the evolution of humanity is developed through seven stages, pulses of Light, with a materialist interpretation—different races—that fit well with the general ideas of her own time. To sort out what may be true and not true in the various esoteric traditions of the Planetary Underworld, there is only one criterion we may apply— empirical evidence.

Generally speaking, it is time to take pains to avoid the "physicality trap" when discussing the future of humankind and the universe. The difficulties, or even catastrophes, in the time ahead will have much less to do with natural disasters than with the social, spiritual, and psychological consequences of the old values coming to an end as a result of a change in consciousness. The constant pull toward physical factors as causative in evolution goes back to the blindfolded Planetary Underworld where the ideal in science was that everything should be reducible to physics. Soon, as we come a little farther into the Galactic Underworld, the science of physics will become subordinated to a unified understanding of the cosmos based on the evolution of consciousness. For such a unification to come about, charting of the oscillations of the World Tree is crucial.

Finally, we have every reason to consider the Galactic Underworld to be identical with what in Christian terminology is called the Apocalypse, the scenario that takes place in the final phase of creation and is metaphorically described in the Book of Revelation. This seems very likely based on the many mentions of the number 7 (fifty-two times) made in Revelation: seven angels, seven spirits, seven trumpets, and so forth. These sequences of seven are described as pulsewise steps in progressions and seem to be identical with the sequences of seven

Days of the different Underworlds that are all going to be completed at the same time, October 28, 2011. Contrary to popular opinion, the Apocalypse is not the end of the world, or at least it is not meant to be. Since the large-scale evolution of human consciousness is entirely determined by the cosmic plan, the Beast cannot win, and so the Book of Revelation ends with a world freed of pain and suffering—the end of the world *as we know it.*

Through an understanding of the Nine Underworlds of creation, the central message of Revelation—that God eventually will defeat evil—not only makes sense but becomes a logical outcome. Because evil is created in the beholder by external projections of dualist frames of consciousness, it will come to an end as these frames are transcended. Of course, it is true that human history and current human life is full of "evil," events with painful consequences for those affected, but it is nonetheless very important to realize that evil and good have no independent existence "out there." Evil acts are only the results of negative spirals of actions generated by projections of dualist perceptions. Despite much Hollywood mythology, there are no forces of good and evil, and paradoxically those who promote such views are contributing to the continued suffering on the planet.

8

● ● ●

The History of the Human Mind

THE MODERN MIND AND ITS TRANSCENDENCE

Because we are now entering a new Underworld and so are climbing to a higher level of the cosmic pyramid, it may be appropriate to reconsider the goals we have set for our individual lives and the strategies we have developed to attain them. In many cases, the goals typical of the Planetary Underworld, such as pursuing a career, may no longer be valid. While many may want to continue with their lives as before, this will be impossible, especially considering the high frequency of change in the Galactic Underworld. Those who deny the change in consciousness that the divine process of creation now brings about—or actually imposes on us—and fail to develop their intuition accordingly will sooner or later find themselves at a dead end. If we are about to climb to the top of the Cosmic

Pyramid, the only realistic goal is to seek for ourselves and all others the cosmic consciousness of the Universal Underworld. All goals that fall short of this will, in a few years' time, be obsolete.

In my own view, the Underworld that we are now entering, the Galactic Underworld, is fundamentally about healing, about creating balance both on a global and an individual scale. It is about unifying the East and the West, intuition and logic, spirit and matter. It is also about unifying body and soul and healing the traumas that prevent us from being fully in the present. To allow this to happen, we must choose paths consistent with the wavelike evolution of this Underworld. Thus we are presented with a choice as to whether we want to integrate or oppose the new energies.

Yet as in all healing, the initial effect may be a deterioration in the condition of the patient. Something parallel may be the result, both on an individual and a collective scale, of the onset of the healing effects of the Galactic Underworld. Such initial impairments sometimes play the role of wake-up calls. For those who have been called, it is very important to have an understanding of the cosmic plan as a basis for their choices. Otherwise negative spirals ending in despair can easily result.

Before going into what a person might do to heal the effects of the imbalances of the individual as well as the global mind (which are inseparable, as they are in holographic resonance), let us briefly examine the nature of the modern mind. The distinction between the mind and the higher Self, two aspects of human existence, is crucial to Eastern philosophies for spiritual evolution and healing. The Self is seen as infinite and eternal, while the mind is limited in both time and space. The mind is a set way of structuring our thoughts, producing a constant stream of chatter, comments, evaluations, comparisons, and judgments. These judgments and evaluations always lead the modern mind to look for something wrong, either in you, in somebody else, or in the situation. The mind cannot rest in the present; instead it oscillates between past and future, always looking for ways to change things. It is not intent on being but on becoming and so leaves no room for peace or satisfaction. This modern mind thus sees no value in sensual experiences as such but seeks to use them only for its own purposes. With the exception of maybe a thousand enlightened people in India, and a handful elsewhere, this mentality now dominates everyone alive.

Based on the knowledge we have extracted from the Mayan calendar, it is apparent that the modern mind is a product of the frames of consciousness of the National and Planetary Underworlds. Through our common resonance with the creation fields of these Underworlds, this dualist and impersonal mind (see fig. 3.11, page 46) is imposed on every human being. Thus the mind that provides the structuring of our thoughts is not personal. Although modalities may exist, we are all essentially subordinated to the same mind.

As part of this condition, it is easy to see that a basic dissatisfaction with our existence must derive from the fact that we are only half. Ever since the beginning of the National Underworld—the Fall—half our experience, that half mediated through the intuitive right-brain hemisphere, has increasingly been filtered out. Yet many, who see this "halfness" in everybody else, think it is a natural, or even eternal, condition and that this is what it is like to be a human being. Because the individual accepts the mind as natural, its halfness will lead to the conclusion that there is something wrong, either with himself or herself or with others. Something always seems to be missing, but it is not obvious what. The dominance of the duality of the National Underworld also leads us to make constant comparisons, judgments, and evaluations of ourselves and others, and thus we are not able to be fully present as complete beings. To the individual the mind is invisible, much as water is to a fish. The mind's response to the halfness that the ruling frame of consciousness has generated is to suggest things for the individual to "become," or to "change" things in the external world. It is not to find peace.

To this dualist foundation has been added the blindfolded consciousness provided by the Planetary Underworld, which has made us look for a remedy, or compensation for our halfness, in the material world. A typical strategy of life in this Underworld has thus been to make a career or to become rich. But this focus on "becoming," to move into the light aspect of the duality, as it were, is just a product of a mind that has been developed in resonance with a specific combination of Underworlds. While we may experience temporary gains from accomplishing such life goals, they will never bring full satisfaction. This is because they are not of the Self and have been generated to serve as compensations for a mind that is not whole. Moreover, this partial mind will not last, since it is always being transformed by the energies of the cosmic plan. Even in those of us who seem to have "made it," there is usually a strong unconscious fear linked to the fact that, with a yin/yang duality that is not permanent, our place in the light can never be taken for granted. Feeling good, being happy, or "living in the light" does not mean the same as being enlightened. It only means that we, at least temporarily, perceive ourselves as being in the yang aspect of the polarity dominating our mind.

The "halfness" of the modern mind, resulting from the underlying duality of the National Underworld, is at the origin of many of the most serious problems facing modern humans. Hierarchies and relationships based on inequality are based on the desire of an individual that is half to become whole by complementing his or her own halfness with the halfness of another individual in the illusion that this will lead to both of them becoming whole. This sense of lack, originating in the halfness, is really the cement of all hierarchies of dominance, whether democratic or not. On a more personal level, it is also the cement of many dysfunctional relationships between individuals. In codependent relationships, the partner is essentially given the impossible task of helping another individual experience wholeness, and as soon as one of them seeks a path toward wholeness, the cement of the relationship starts to crumble.

The halfness of the dualist frame of consciousness creates a more or less permanent sense of lack in people. This generates ideas on a larger scale related to a perceived need for "growth." We see this in population growth in poorer countries, where the driving force is not always the desire for children as such, but the striving of parents to escape the sense of lack generated by the experience of halfness. We also see its result in the infatuation of many modern societies with "economic growth," as if it were an end in itself. This drive for an abstract growth of the economy, regardless of actual human needs, is a very dangerous aspect of the modern mind. The ambition of some to alleviate the sense of halfness through idealizing "growth" is what threatens to destroy the ecological system of the earth and humankind's place in it.

If we have been wearing such glasses provided by the mind, especially without even knowing it, it is no wonder that the world looks the way it does. Because everything we do is ultimately an expression of the mind, all human-made creations, intellectual as well as material, have been made with these glasses on. Thus all human creations are conditioned by the limits of the mind of the person who brought them forth. From this halfness, we may also understand why there has been so much dissatisfaction and fear in human life ever since the Fall almost 5,200 tun ago. Yet in the understanding of all this there also lies hope. If everything is of the mind, and a mind that is whole is now about to be created, we may be healed, both individually and as a species, if we are only open to the new light of the Galactic Underworld. The world will change for the better if we can learn to surf on these new waves.

Only in the most recent decades have Westerners begun to recognize that there may be such a path to a fuller life, and that steps toward deeper satisfaction may be taken by loosening the grip of the modern mind. The first such influence in the West may have been Zen meditation, which, with its impenetrable koans, aims to short-circuit the mind. Transcendental Meditation, Silva Mind Control, est, Hare Krishna, t'ai chi, yoga, and countless other approaches have followed. Many meditation techniques have been introduced, while others have become obsolete. It is not my purpose here to compare or discuss the applications or effects of these diverse movements and practices. What they have in common is that, in one way or another, they seek to loosen, if only temporarily, the grip of the modern mind, either by transcending or short-circuiting it. Even a temporary experience of this kind may give the individual a spiritual sense of the Self. All in all, these movements have developed techniques that can prepare us for a frame of consciousness that is whole and balanced.

What few people realize, however, is that the very phenomenon of people in the West and in Europe starting to learn meditation techniques was a reflection of an energy shift in the divine plan, as may be understood from the Mayan calendar. This goes back to the beginning of the Fourth World of the Planetary Underworld in 1947 and to the initial change in the relationships between East and West and between brain hemispheres that this created. This Fourth World is a general influence that has prepared for the much deeper change in consciousness that is now being developed by the Galactic Underworld itself, where the light in the yin/yang duality shines on the right-brain hemisphere. In the East the effects of the onset of this Fourth World were evident in the independence of India (1947), Indonesia (1948), and China (People's Republic, 1949). While initially these nations were fragile and economically weak, the new World led to the liberation of about half the world's population from direct Western control and the beginning of the influence of the East on global civilization.

In the West, the onset of this Fourth World was most immediately visible in a number of innovations, in early embryonic forms, that pulled down the barriers between material and spiritual, body and mind, and so on, which would later lead toward a more holistic worldview. These more holistic views emerged in response to the difficulties that had became associated with modern life. Hans Selye, the pioneer of stress research, for instance, who established a mind-body connection in the creation of stress, first used the term *stress* in 1946. Although the idea that psychological phenomena may contribute to physiological stress reactions may seem self-evident to today's practitioners of alternative, or even mainstream, medicine, they were not regarded as such at the time. This represented the first major recognition that a psychological condition may cause an adverse state of the body. And so as the Fourth World began, we saw the first embryonic signs of the healing of the Cartesian split between body and soul that had been imposed on humankind as Ometeotl/Omecinatl began his/her rule in the National Underworld in 1617.

That this mind-body connection was first recognized in the field of stress is not an accident. As a consequence of the beginning of the Fourth World, the mind started to be influenced by the higher tun frequency of change that only now in the Galactic Underworld is fully developed. This meant an approximately twentyfold rise in the frequency of change compared to what had dominated the Planetary Underworld. As a result, stress became a much-discussed adverse aspect of modern life. From the beginning of the Fourth World on, the link between the general life situation of a person, the frequency of change, and his or her state of health could no longer be ignored, although no one at the time suggested that the Mayan calendar had anything to say about it.

What, then, is *stress?* It is not really a product of modern life. People in earlier times did not have all the conveniences that we now have at our disposal, and they had to work much longer hours. Yet there was no talk of stress. Stress is a product of the modern mind and reflects the conflicting influences on the individual of two different frames of consciousness linked to two different Underworlds. As the consciousness of the Galactic Underworld made itself felt to a low degree around 1947, and then much more markedly as it actually began in 1999, the high-frequency spiritual pulses of light started to be transmitted to the right-brain hemisphere. In unknowing people the influence of these pulses created a conflict with the already dominating materialist values of the Planetary Underworld. This resulted in an experience of stress, which sometimes has physiological consequences.

Today, burnout has become a major adverse condition, and the number of afflicted is skyrocketing. The conflict between the rapidly developing higher consciousness of the Galactic Underworld and the society around us, which seems to be dominated by the values of the Planetary Underworld, can now easily lead to a state of paralysis. Burnout is also a sign that many people (subconsciously) now desire to transcend the consciousness of the Planetary Underworld. The body seems to be telling the individual that it is not being aligned with the divine plan and refuses to go on. Those afflicted are not the lazy, but rather the compassionate—individuals who cannot shield themselves from the effects of the high-frequency pulses of the new Underworld. For many the burnout condition may be a wake-up call, an initial deterioration resulting from incoming healing energies, that leads them to explore the spiritual dimensions of reality.

It is no wonder that when the meditation techniques of the East began to be introduced in the West, initially through Zen but later notably through Transcendental Meditation as taught by Maharishi Mahesh Yogi, they addressed the condition of stress. Why? Because stress is a condition of the mind. One way of alleviating its effects is to transcend the grip of the modern mind that had caused the condition in the first place; and meditation, which makes the individual aware of the constraints of the mind, serves exactly this purpose.

It has been shown that the activity in a certain area of the left frontal lobe of the brain is decreased in a meditative state. It turns out that this very area is crucial for the human experience of physical being, because it integrates nerve impulses from many parts of the body. As a result of the decreased activity in this area of the brain among persons who practice meditation, they will identify more strongly with the spiritual aspects of their being and enhance their intuitive faculties. It is noteworthy that the area that creates the experience of physicality is in the left frontal lobe, as we can then expect that its role will be balanced by the emerging light on the right brain. In this way, we can understand that the change in consciousness brought about by the Fourth World meant a deterioration—stress, and later burnout—but also remedies, techniques from the East favoring the information mediated through the right brain, which generates a more spiritual worldview.

But the new energies of the Fourth World, in providing light to the right-brain hemisphere, also had a more general influence in the West. This was the baby-boomers' rebellion, or the hippie movement, in the generation that was born around the year 1947, whose members had been exposed to the general influence of this World throughout their entire upbringing. This rebellion, which at the time was seen primarily as a generational conflict, was really a conflict between Worlds in the Mayan sense. As those influenced by the Fourth World reached maturity around 1967–1968, they rebelled against the idea of making a career, experimented with mind-expanding drugs, and so forth. From Berkeley to Beijing, with notable Paris in between, different expressions of an antiauthoritarian movement swept the world with an unprecedented synchronicity in the year 1968. More than ever before the effects of a Mayan calendar cycle, the Fourth World, could be seen to have independent and synchronistic repercussions worldwide. And while we may today regard the hippie movement, or the more political rebellions in other parts of the world, as naive or unrealistic, its aims reflected an early dissatisfaction with the constrictions imposed by the modern mind and an unfulfilled desire for enlightenment.

In general, in the West the influence of the Fourth World strengthened a way of being based on the intuitive brain hemisphere. It brought ideals of peace, wholeness, mysticism, and so forth. It also provided the general background for the women's movement, the ecology movement, and the New Age movement, and an appreciation of the contributions of indigenous cultures. These were all results of the initial unification of brain halves. Although the movement was soon put down by the forces of the left-brain hemisphere and the Planetary Underworld, it was nevertheless a prelude to the phenomena that, in an entirely new form, will be expressed in the Galactic Underworld.

From the description of the cosmic plan by the Mayan calendar, we also have an answer to why the emergence of the hippie movement was one of the later signs predicted in the Hopi Prophecy. The shamans or prophets of this people were able to be in resonance with the energies of the cosmic time plan and to "see" the directions in which it would lead. It led them to see a shift toward the aspects of the cosmos mediated through the right-brain hemisphere. As a step away from Koyaanisqatsi, they were also able to foresee the coming of the "flower children." The only reason that such prophecies, precognitions, premonitions, and miraculously fulfilled ideas are possible is that there exists a cosmic plan. Unless there was a strict time plan for the evolution of the cosmos, events would happen at random and would be entirely unpredictable.

That the mind has a history defined by the Mayan calendar may be one of the most important discoveries of all time, and its implications should be fully recognized. It should not be confused with the idea that human thoughts and ideas have a history, which is trivial and more or less obvious to everyone. To say that the mind has a history is instead to say that the mental structure organizing human thoughts and ideas changes according to an exact time schedule. Thus what is transmitted from the otherworldly domain through the human brain also has a history. This discovery is fundamentally a reason for hope. The divine process of creation is designed to lead the human mind to an inner state of peace and hence to an absence of external conflicts.

UNITY OF MIND AND EXTERNAL PEACE

In general, the particular Underworld by which human beings are most immediately dominated has a profound influence on whether their outlook on life is dualist or unitary and so in turn to what extent they will engage in conflicts or peace; this was briefly discussed in relation to the Planetary Round of Light. What also affects whether there is war or peace is to what extent people are aware that they are dominated by a dualist or a unitary mind. If we are aware of the nature of our mind we at least have some choice about it. The reason humanity has lost faith in the idea of a millennium of peace is that it has become blind to the existence of a cosmic plan, according to which our minds change. Unlike the calendars based on physical cycles, the Mayan calendar tells us that at the end of creation we will be ruled by a unitary mind, or, as the Hopi say: one world, one nation, under the Creator. Thus the road to peace is largely one of transforming our minds to a unitary enlightened frame of consciousness, and, by taking such a path, deepening our realization that we are all one. Spreading the knowledge about the cosmic time plan and the use of the Mayan calendar are then also means of creating peace.

From the perspective of the Mayan calendar, there is a deadline by which peace is to be attained—October 28, 2011—and much of the current defeatism, or sometimes illusory ideas, about the prospects of creating peace on earth actually derive from a lack of understanding of the cosmic time plan. If peace has not been attained at the completion of the plan, the human species on this planet will not have reached the level of consciousness it was meant to and so will probably destroy itself.

From the perspective of the cosmic plan, there is thus no such thing as a set "human nature." How human beings think and act is instead determined by the particular Heaven and Underworld that rule them at the time they happen to live. Wars are primarily the result of conflicts between groups of people dominated by different energies and levels of consciousness. Whenever there is a significant change in energy because of a cycle shift, the balance between the social structures and hierarchies that have been based on these levels of consciousness is affected. Conflicts, armed or not, almost always surface when the energies shift.

This is why there is still no peace in the world. The overwhelming majority of humanity is still dominated by dualist frames of consciousness. What may be worse, in the current Galactic Underworld the dualities are also shifting at such a high frequency that it becomes a necessity to transcend them to avoid worsening the situation. It is the changing dualities dominating the collective mind that create wars and armed conflicts. There are countless examples from human history illustrating this point. Consider, for instance, the pattern of violent movements described in chapter 3. Wars are not usually fought for the reasons outlined by their instigators or participants; rather they are fought because new Heavens and Underworlds are beginning to dominate the human mind and so generate shifts in earthly power. Those who are surfing the waves of an emerging Underworld will usually come out as victors, while those who surfed on the old will be defeated. The outcome has very little to do with who is "right" or "wrong" or who is "good" or "evil." It has more to do with who, at any given moment, has the wind of the divine process of creation at his back. Thus, throughout history, human beings have been puppets of spiritual cosmic forces, puppets of the energies of Tezcatlipoca, Quetzalcoatl, and so on; and even more profoundly, of the yin/yang polarity of the particular Underworld by which they happen to be dominated.

Usually, with a climb to a new Underworld a new set of values is introduced that creates entirely new types of conflicts. In the Regional Underworld, organized wars were virtually nonexistent, and, as related in the Book of Genesis, the perception of "good" and "evil" first came with the Fall. The dualist way of looking at the world then emerged because of the establishment of the yin/yang polarity by the World Tree in 3115 B.C.E., and we know for a fact that around this time the first organized warfare began. From this point on, wars have been fought between "us" (placed in the compartment of light, the "good" ones) and "them" (placed in the compartment of darkness, the "evil" ones). Once the human perception of evil is projected onto the external reality, evil will easily be manifested as a reflection.

In the National Underworld, wars were mostly perceived as fought for the glory of kings and nations or in the names of some of the historic religions. Let me take the case of my native Sweden as an example. Its armies in the Thirty Years' War saw themselves as fighting for the glory of their king and the survival of the Protestant creed. In reality they were propelled by the new duality introduced as the Thirteenth Heaven of the National Underworld began. They were the puppets of the energy of Ometeotl/Omecinatl, and so at this time they had the wind of history in their sails. The duality introduced by Ometeotl/Omecinatl then gradually created four hundred years of Western dominance, expressed primarily through the phenomena of capitalism, science, Protestantism, and the modern nation-state.

At shifts between ruling deities, those that have the new energy at their backs experience the shift as divine inspiration for their missions. At the time, the Swedes surely thought God was with them, favored their church, and was against the papacy. Yet they were manifesting in physical reality what in one way or another was meant to happen, essentially for no other reason than that they were living in a special location in relation to the World Tree. Similarly, Ometeotl/Omecinatl gave the wind to the Pilgrims who landed in Massachusetts in 1620, from where they would gradually win the West. Winners or conquerors in wars invoke national, political, racial, or religious superiority as explanations for their victories, but what is really operating is a temporary spiritual wind at the back from one of the Bacabs (see fig. 3.1, page 35), the directional gods of the Maya.

The new energies of the Planetary Underworld led the United States to world dominance. This Underworld led this country from the Albany Congress in 1754, when a union was first conceived, to its emergence as the only superpower as the Soviet Union collapsed a month before the katun shift in 1992. Its current role in the world is nothing but a product of these energies. With the blindfolded consciousness of this Underworld, wars were fought primarily for control over the physical reality, natural resources, and so forth. And while some may think such motives are the only causes of wars, they are really just expressions of the particular frame of consciousness generated by this Underworld, a frame that, as we have seen, allows us to see only the material aspects of reality and places the spiritual "somewhere else."

Until recently, wars with religious motives had essentially disappeared in the Planetary Underworld. Yet in the now emerging Galactic Underworld we have seen a surge of Muslim suicide bombers (although such acts, in fact, are against the Islamic creed). These acts of violence are not carried out for the glory of a nation or a king, as was the case in the National Underworld. Nor are they fought for control over natural resources. In the Galactic Underworld a new motive for violent acts is established; to weaken the West and its material dominance with no possible individual gains for the perpetrators. Ultimately, all conflicts are the results of changing ways of perceiving reality, and these are the initial expressions of the consciousness favoring the East that were developed by this Underworld. We may note, however, that the Easterners who violently clash with the West are those individuals who have a strong left-brain dominance; because the centers for writing and reading are located in the left brain, people with fundamentalist belief in written scripture are typically left-brain dominant.

Fundamentally, however, the Galactic Underworld is about healing the world. This is obvious if you consider that the light on the brain hemisphere that has dominated humankind for the last five thousand years may now be complemented with light on the other. What the Galactic Underworld really brings is not a set of new "things," but a new way of perceiving reality generated by a duality favoring the Eastern Hemisphere. Inherent in the Galactic Underworld are thus possible conflicts of views between the East and the West. Yet, of course, the yin and the yang of creation are not sharply separated. There is some yang in the yin and vice versa, so the planetary midline is very far from being a sharp line of division. Individuals and collectives in both the West and the East will be influenced by the new pulses of light favoring their right-brain hemispheres. To the extent that we, collectively speaking, are able to embrace the holistic perspective of the right-brain hemisphere, the dualities will even out, and the rift in the world will be healed.

Surely, then, the United States, representing the energy of the West, is about to lose its leading role in the world. Although it may well for some time retain a military, technological, and media superiority, it no longer inspires the rest of the world. In the now-emerging Underworld, material superiority no longer confers true leadership, and there is no worldly force that in the long run can withstand the winds of history generated by the cosmic plan. In the Galactic Underworld, spiritual inspiration will instead come from the East, because at the present time this is the only direction from which balance and wholeness could come. The main spiritual leadership will initially come from India, but at later stages China and Russia also become important in this regard. Japan, which for some time has been regarded more or less as a Western nation because of its advanced technological level, will soon come to be regarded as one among many Asian nations. At least culturally, the United States will increasingly be seen as distinct from Europe, and there will be a sustained movement in the direction of a divorce. As the World Tree generates distance between the two, Europe will come to play an increasingly important role as a coordinator and transmitter of the new energies. This is not to say that Europe will again take on a dominating role as it did in the colonial era; the whole process of the Galactic Underworld is about putting an end to dominance and imbalance. As it draws to a close, there will be no global leader, and the world will be prepared for true brotherhood and equality in all respects, including economic. For the enlightened person who knows the cosmos as her homeland, patriotism and nationalism are alien things.

It is not that the cosmic energies will punish those who go against the cosmic plan, but rather that if we go against the long-term spiritual winds, we can be sure we will not succeed. If we insist on trying something that is counter to the cosmic plan, such as holding on to the world order established by the Planetary Underworld, there is certain to be a backlash. The way the cosmos operates, such a backlash may come in any number of different ways, sometimes in ways that seem totally unrelated to the process that caused it. If there are catastrophes or wars, the best question to ask is: What are we doing to make the cosmic plan lash back at us? Considering the high number of people today who are focused only on the material values of the Planetary Underworld and who do not even know there is a deadline by which the purpose of the cosmic plan must be fulfilled, there is every reason to believe that the cosmic plan will lash back, possibly on a very large scale. In one way or another, the cosmic plan paves the way for the direction creation is meant to take, and it is outside human power to change this. The more we align ourselves with the ultimate purpose of the cosmos, however, the fewer conflicts linked to its development there will be. This is a fundamental principle that requires from all of us serious and rigorous probing and discussion, based on the Mayan calendar and empirical evidence from all aspects of evolution, of the purpose of the cosmic plan.

A person with a unitary outlook on life ("we are all one") realizes that if there is not peace on earth, every single individual has something to do with it. From such a point of departure every one of us may contribute to creating peace, also in an external sense, through our commitments to attain the enlightened state. The Mayan calendar teaches us that all external manifestations are products of the mind, and so if we are able to peacefully and constructively adapt to the change in consciousness that is currently imposed on us by the cosmic plan, we may be able to contribute to peace. The dualist outlook on life, on the other hand, creates the idea that there really is "good" and "evil" out there. And with such an outlook the only possible strategy for peace is to eradicate the "evil." But then again, if you do, what is your basis for including yourself among the "good"? A dualist outlook never leads to peace, as it does not address what created the violence in the first place: the dualist mind.

It is not an accident that the most conflict-ridden area today, ranging from Egypt to India via Israel and the Palestinian territories, Iraq, Afghanistan, and Pakistan, is exactly the area where the world's oldest civilizations—Egypt, Sumer, Persia, and the Indus Valley—first saw the light of the day. This is the area that for the longest time has been dominated by a dualist frame of consciousness and where a strongly patriarchal mentality has consequently been fostered. At the present time, with the onset of the new energies of the Galactic Underworld, it is not surprising that this region is deeply affected.

The good thing about the present situation is that we are living in an Underworld that is designed step by step to generate a more unitary outlook on life, and so we need to let the two opposing dualities merge rather than engage in their respective dominance games. This is not about changing negative thinking to positive or engaging in wishful thinking about peace, but about something more difficult, which is to altogether transcend the yin/yang polarity through which we experience reality. The road to peace involves creating peace within ourselves, between our own brain hemispheres, and then projecting peace externally. In this way, a peace can be created that is not just about preventing and managing conflicts in the external world.

Nowhere will the effects of this peace be more evident than in Jerusalem, the holy city of the three Abrahamic religions. The beginning of the Fourth World in 1947 brought not only the independence of the major Asian nations and the cold war; it also planted a seed of conflict between East and West, the United Nations' partition of Palestine (1947), and the founding of the state of Israel in 1948, which since then has been an area of conflict. In the Middle East people are still strongly attached to the religious thoughts that were generated by the consciousness of Tlacolteotl (Third Day, Moses), Quetzalcoatl (Fifth Day, Christianity) and Tezcatlipoca (Fifth Night, Islam) of the National Underworld, and so are dominated by the filters of these deities. We find Jews and Christians on the same side because their religions were both generated by Days, while Islam reflects the energy of a Night. This is a conflict between the thoughts of Days and the thoughts of Nights. It could also be seen as a struggle among the deities ruling the mind who are using human beings as their pawns. Such a perspective may seem provocative, but it is closer to the truth than many would think.

Given the existence of these conflicting religions in Jerusalem, why does the Book of Revelation describe the bright world that will appear at the end of creation as the New Jerusalem, when this is today probably the place in the world where people are the most pessimistic about the prospects of peace? If we are not talking only about a managed peace agreement, but peace in the sense of a deep mutual longing for unity, the prospects would seem even worse. Put in other terms, however, the reason that there is no peace in Jerusalem is that three different religions and two nationalities with origins in a dualist Underworld are entrenched there. If there is peace in Jerusalem, there is surely also peace in the rest of the world, because Jerusalem is the place where the dualities of the lower Underworlds will be the most difficult to transcend. And if it is the most difficult place, the only way peace can come about there is if the whole world transcends these dualities out of a longing for unity and enlightenment and then brings this to Jerusalem. Thus if the world attains the new frame of consciousness the creation of the New Jerusalem will be included, and from this the prophecy in the Bible may be understood. The New Jerusalem is really the New Enlightened Mind.

This shifts the perspective somewhat. Even if outsiders cannot easily intervene in the Israeli-Palestinian conflict, the whole world may be able to contribute to peace there by focusing in their own lives and in their own regions and in all the conflicts in the world on a path toward attaining the consciousness of the Universal Underworld. If the attainment of a cosmic consciousness of peace is recognized as a real possibility in the rest of the world, it will be recognized in Jerusalem too. The situation in Jerusalem is a global indicator of the lack of transcendence of the modern mind and of how little humanity has progressed on the path toward enlightenment. When we have a New Jerusalem, we also have a New Enlightened Planet, because only inasmuch as we have all become enlightened and transcended our dualist minds will there be a New Jerusalem.

WHOLENESS AND LOVE: THE HEALING EFFECTS OF THE GALACTIC UNDERWORLD

The main path provided by the Galactic Underworld is thus toward wholeness and love, and this is charted at the level of consciousness. As not everyone may want to join the climb to higher levels of consciousness, however, there are also possibilities for conflicts and events that may be seen as negative. On a personal level, in their relationships to themselves and to others, people are being deeply affected by the onset of the new Underworld. To see how this Underworld will bring love and wholeness, it is best to begin by looking at relationships within the generation that was shaped by the Fourth World.

This World shook the traditional family structure deeply and broke it down in significant ways. Even if the Fourth World generation that broke out in the worldwide rebellion of 1968 eventually returned to the hierarchical structures of society at large, the new light on the right-brain hemisphere did create a very different and more egalitarian family structure. A profound change took place in the relationships between men and women. Very different ideals also developed for the role of children. Anyone who has read the books of Alice Miller, or has direct experience growing up in the early twentieth century, can attest to the fact that children in this time were not seen to have great value in their own right. Even if there were exceptions, many people—even in more recent generations—struggle in one way or another for the rest of their lives to heal the effects of occurrences in their early childhoods that have exerted a negative influence.

What does our current perspective of different Underworlds ruling the mind of humanity have to say about this? For one thing, family and relationships are ruled by the same mind and the same dualities as the society at large. The typical family structure that has existed ever since the Fall has been based on the same hierarchical structure, in which the husband was traditionally a ruler or breadwinner. It is in the very nature of the dualist mind of the National Underworld to generate hierarchies, inequalities between people, and in the family children and women were in the yin category. Gradually, however, after the onset of the Fourth World of the Planetary Underworld, this family structure has become increasingly dysfunctional. Today the typical four-member family, ruled by a single head, is more of an archetypal memory invoked in commercials than by a current reality.

With the Fourth World, the beginning of a balance between the left and right-brain hemispheres gave rise to the beginning of a balance between men and women and between adults and children. The changing balance has also generated many divorces, partly because many were not able or willing to adjust to it, and this has resulted in a very diverse family structure. This change in consciousness has certainly fostered equality, but it is also a good illustration of how there may be complex consequences of the influx of healing energies. On one hand, the new balance may favor wholeness, but on the other the breakdown of the social structures built on the dualist mind has also caused divorces that in most cases are not easy. It is likely that the Galactic Underworld will intensify the turmoil of relationships. The rapidly shifting energies between Days and Nights every tun of the Galactic Underworld are likely to shake all relationships strongly, and many of those that are based on dominance can be expected to break up. Again, as in the case of burnout, it will be those who are the most sensitive to the incoming new light who will be first to be affected.

But our relationships go through phases, and we should look more deeply at how the evolution of consciousness influences our individual development throughout life. Children enter the world with a unity with All That Is, which initially takes the form of the mother. In the development of the individual, this unity corresponds to having the consciousness of the Regional Underworld, in which people similarly lived in a unity with All That Is. This probably took the form of Mother Nature or Mother Earth, although it was likely so obvious at the time that there was no name for it. Then, in a series of steps, essentially between the ages of three and seven years, children acquire the consciousness corresponding to the National Underworld. They become aware that they are separate from their mothers and learn all kinds of distinctions, such as between "right" and "wrong," as well as how to read and write. As part of our individual climbs of the cosmic pyramid, in other words, we acquire a dualist mind. Around the age of fifteen to eighteen years, people then begin to acquire more of the consciousness of the Planetary Underworld—they learn to handle machines such as cars, for example, which are typical of this Underworld (invented in 1884, its midpoint). After this the individual is supposedly ready to enter society at large, which currently operates according to the materialist values of this Underworld.

Today, as a result of the onset of the Galactic Underworld, changes are taking place in the ages at which human beings in their individual development come into resonance with the consciousness levels of the different Underworlds. Many children, of course, enter the information-technology reality of this Galactic Underworld at a very early age. We now also hear about the preteen age, especially among girls; and some people are talking about so-called Indigo children, who are said to have an already very advanced resonance with the cosmos and "psychic" abilities. (Indigo supposedly refers to a new chakra, which has presumably been developed by the Galactic Underworld). All in all, it seems that those growing up today are finding ways to enter the climb to the higher Underworlds at an earlier age in their personal development than previously. These children simply refuse to waste much time in the lower ones.

Returning to the topic of healing, the point when most traumas needing healing seem to occur is in early childhood, at about the time when the dualist mind comes to dominate us. Healing is then often about repairing the effects of this duality. Many people, at later times in their lives, find themselves missing the loss of childhood innocence (before having entered the consciousness of the dualist National Underworld). Children aged two to seven years who have not been abjectly mistreated often even invent traumas to explain why at some point they became separate. But in many other cases the traumas that caused the separation were very real. Because our individual development parallels humanity's climb of the cosmic pyramid, we are at this point personally expelled from the Garden of Eden (no wonder that many have believed in an "original sin" common to all humankind).

In one way or another, the imposition of the dualist mind creates separation, a separation that we now, as part of our path toward enlightenment in the last two Underworlds, will need to heal. The traumatic event in our early childhood really did cause pain at the time, but the only reason that a trauma of early childhood still influences us is that it is being used by the mind. The modern mind, we may recall, lets our past dominate our present.

Before the change in consciousness brought by the Fourth World, the more or less automatic response of an adult remembering childhood trauma was to assume the perspective of the perpetrator—usually the parent: "Getting spanked never hurt me." When you are completely dominated by a dualist frame of mind, as people still were in the Third World, the easiest way of relating to a trauma is to identify with the authority figure who inflicted it and place him or her in the yang (light) category of the mind, and so make both you and the authority figure "right." Sexual abuse of minors, for instance, very rarely came to the surface and certainly was not a matter of public discussion. Only in the latter part of the Fourth World was this allowed to pass through the dualist filters. Before 1947, it is doubtful whether therapies for psychological healing ever generated any deeper results because the issue of the dominance of the dualist mind was not addressed. Psychoanalytical techniques, for instance, were only rarely able to go back to a time much earlier than the child's ability to make distinctions and speak. The wholeness of the pre-Fall level of consciousness—which is what makes healing possible—was never reached.

Around 1947, however, therapies were developed that were not merely mental. In the late 1940s Fritz Perls pioneered Gestalt therapy, which had as its key concepts "Here and now" and emphasized emotive processes. Stanislas Grof, working with LSD therapies a few years later, found through his research that events taking place very early, even at birth, could have traumatic effects; he also sought to integrate Eastern philosophies in his therapy. An opening was thereby created for methods such as primal therapy, neurolinguistic programming (NLP), rebirthing, and countless others whose emphasis was on emotional integration in the present. One of the more powerful therapies today coming out of the Galactic Underworld is Brandon Bays's Journey (actually a journey to the Regional Underworld). As a common thread, all these therapies have the goal to heal traumas of the "inner child," or even previous lives, essentially through accessing levels of our consciousness that are pre-Fall, early in the development of the individual, and then allowing a new start from there.

From our current perspective, this new wave of healing techniques was generated by the onset of a World of light shining on the right-brain hemisphere. The idea of healing one's inner child was not just a good idea that popped up out of nowhere. It had an origin in the larger cosmic plan, with the purpose of healing humanity in preparation for attaining the enlightened state. Healing is essentially about going back to a point in our lives prior to the time that the dualist mind came to dominate. As this mind is preoccupied with making judgments and evaluations, it creates a universe of "right" and "wrong"; it must be "right" about the past and its traumatic events, and so one is unable to forgive either others or oneself. Forgiveness mostly requires transcending the control of the modern mind and accessing, if only temporarily, the healing powers available in unitary Underworlds.

In two significant ways, the qualities of the right-brain hemisphere support forgiveness. First, they allow you to trust your own intuition, which is of the right-brain hemisphere. Trusting your intuition is really about trusting yourself, your whole being, sometimes in the face of seemingly logical arguments of others and the constant chattering of your own mind. Second, they balance the dominance of the left-brain hemisphere, which organizes events in your life linearly in time according to a cause-effect scheme. In a wider perspective, a traumatic event may not have had the place in the cause-effect scheme at which the left-brain hemisphere is fixed, and in a wider perspective things might have a different meaning. Hence, healing is possible if we can transcend the linear time of the left brain. Such transcendence is exactly what the Galactic Underworld will favor. A relevant question to ask regarding healing practices will then be: Do they serve my climb of the Cosmic Pyramid?

Not everyone will be thrilled by the increased possibilities for healing that this Underworld will bring. Many individuals in families, as well as in society at large, would rather enjoy the benefits of dominance and inequalities than allow everyone to be healed. In some cases the consequences for someone seeking healing may be a separation from others who may not want to join the climb. Some will choose to stay in the grip of the lower Underworlds, mostly without even being aware of it, while others will take on the challenge to climb to higher levels, even in the face of the many limiting thoughts of our current mind.

Thus, as is the case with the world at large, it is simplistic to say that the Galactic Underworld is "good." It depends on what your life's goal is. This Underworld, however, is designed to bring an end to the dominance of one person over another and to heal the rift caused by the Fall. To seek individual healing in this sense is not an egotistical endeavor. At the current time, every individual who sets a path toward, or attains, enlightenment is a help to the rest of humanity. Individual healing helps free humanity as a whole by disposing of "negative karma" that may have accumulated over several generations in a family, or, in fact, in all of humanity, since about the time that the dualist mind first came to rule about five thousand years ago.

There is also a judgmental side to the dualist mind that inhibits the flow of love. If you and everyone else are constantly judged by this mind, how can you allow yourself and others to fully be who you and they are? How can you fully accept and love someone else, let alone yourself, with a dualist mind that was designed to be separate and causes separations? Can a person who is only half love herself fully? How can you fully love yourself when your self-worth depends on being in the yang category of the mind? Even if you have attained that position in your own mind, or even the collective mind, you will have to live with the fear of losing it. Even if we sometimes rise above the surface to see something else, most of us are swimming in muddy waters much of the time. Only wholeness—a mind that is truly unified and whole—can bring true love. Only the whole person who is allowed to be herself completely and does not need to shut out any aspect of herself out of fear of judgment by the mind can truly love herself. Only the person who has freed himself of the dominance of the left-brain hemisphere can allow the whole person, and the whole world, to be. The truth is that the dualist mind still ruling at the present time does not bring love.

The purpose here is not to belittle what anyone may have accomplished in terms of loving relationships, but only to say that if there are difficulties, these may not be as personal as is commonly thought. The message is also to point out that for those willing to make the climb, there may be a wholeness available at the top that is unknown at present. The consequence might be that everyone attempting the climb may seek to heal themselves and their relationships through accessing the light of the Galactic Underworld. Part of the climb involves developing your intuition, which is to put the In Lak'ech philosophy into practice by recognizing the unity of all things. At present, there are many approaches available along those lines, such as courses that teach how to talk with animals, plants, nature spirits, and so forth. Developing telepathy and intuition will prepare us for the unity with All That Is that will dominate the Universal Underworld.

Another aspect of the climb of the Cosmic Pyramid is forging a path of action in service to the enlightenment of humanity. Love and wholeness require including everyone and jointly working with others toward this goal. The Mayan calendar means there is an overall purpose to the cosmic time plan, and so we all potentially have missions, paths of action, in service to the rest of humanity. Although not everyone will necessarily have a specific God-given individual mission to accomplish in his or her life, it is always possible to create one in service to the cosmic plan, once you feel confident about its purpose. Such a path of action, however, needs to be developed within true time frames for its realization, which can only be provided by the Mayan calendar. Many projects have a kind of cosmic urgency that is not apparent in the absence of the perspective provided by the calendar. Your path of action may be crucial not only for your own climb of the cosmic pyramid, but also for others wanting to do the same. The latter's success will enhance the chances of your own.

DESTINY AND CHOICE

Because a blindfolded frame of consciousness has come to dominate the modern world, many have gained an illusory sense of freedom. In reality, as we may now begin to see, human life, both individually and collectively, is fundamentally conditioned by the energies of a divine plan. Since in the modern world, and especially in the West, we seem to be presented with choices all the time, we may easily miss the fact that these choices have been generated within the constraints of the consciousness of the Planetary Underworld. Because of the materialist perception this Underworld has generated, freedom is often equated with the freedom to choose among different things to buy. Not only is this a very limited definition of freedom, it is also largely irrelevant when it comes to fulfilling our true destiny on this planet: climbing the cosmic pyramid.

The existence of a cosmic time plan thus raises critical questions regarding the free will of human beings and their roles as cocreators of the plan. The overall perspective presented here—that human beings have essentially been like puppets of the energies of the cosmic plan, with strong limitations set by these energies on their freedom to act and think—obviously runs counter to a currently very popular idea: that human beings have the power to create anything they like. In such a view, a divine time plan for the evolution of consciousness does not exist, at least not one that humans would not be able to set aside with the power of their thoughts or visualizations.

The idea that there are no limits to human creativity has itself been generated by the very consciousness of the Planetary Underworld. The denial of a cosmic plan can be traced back to the French Revolution, but more immediately to the Darwinist illusion in the mid-nineteenth century. This denial has generated an artificial sense of freedom and made human beings feel they are all-powerful. After all, if it is true, as the neo-Darwinists say, that human beings have come into existence by accidental mutations in DNA, then we would not have any relationship of fundamental importance to the Creator. He/she would then not be our Creator, and we would not be here to manifest his/her plan. Consequently, we would also be completely free to create the world any way we like. The problem with this view is that, based on what we know from the Mayan calendar, it is not consistent with the truth. To some, the recognition that we are created beings, living our lives within the constraints of the evolution of consciousness defined by the cosmic time plan, may come as a surprise.

The truth now emerging is that the energies described by the Mayan calendar define the limits of human creativity at any given time. These energies set the rules for how life is lived in any given combination of Heaven and Underworld. Yet regardless of the energies that temporarily rule the cosmic time plan, they are first and foremost subordinated to its end result of enlightenment, and no matter what we think or visualize we cannot do anything to change this fact. The cosmic plan as such is beyond the range of human manipulation. Moreover, it is fairly safe to say that if human thoughts or actions are not aligned with its intended end result, or if they counteract it, the cosmic energies will in one way or another generate a backlash.

That the cosmic time plan sets limits for human creativity at any given time does not mean that thoughts are not creative or that prayers or visualizations do not work. It seems that communication with the divine, through prayers, visualizations, and other means, may indeed help an individual or a collective create what it desires. But there are limits to what can be created at any given time, and these limits are defined by the frames of the cosmic time plan. To make things a bit more complicated, the limits also vary between different cycles. Thus, for instance, if egotism has been more or less an integral part of the human being in the National Underworld, and the accumulation of material wealth has been quite consistent with the Planetary Underworld, this is now about to be reversed. It will not be consistent with the energies of the Galactic and Universal Underworlds to accumulate material wealth for mere egotistical reasons or to seek control of others. As I describe later, in the now developing yin/yang duality, egotists will ultimately become an endangered species. This will not be in response to moralist sermons. It is simply because this is the way that, on the level of consciousness, the cosmic plan is designed.

But if we are all more or less like puppets of the cosmic energies, what value is there in knowing about them? If the course of events is predetermined by the cosmic plan, what difference is made by knowledge of this? After all, people have been puppets of the divine process of creation for thousands of years and seem to have survived anyway.

The answer is that understanding the cosmic plan is part of our own evolution into cocreators with the Divine. To be cocreators, we need to know what the canvas we are to work on looks like and what its format is. If we choose to surf on the cosmic plan, we had better know what it is. This matter of cocreation may not have been important in lower Underworlds, because these were not designed to make human beings into cocreators in the full sense of the word. But this is exactly what the Galactic Underworld is about. As we climb the cosmic pyramid, we are increasingly being created, and creating ourselves, into images of the Divine. If we climb, this will be the outcome. Part of the process of becoming a cocreator is to know the cosmic time plan. To surf on its waves, and set individual goals consistent with its intended end result, will be much more difficult for those who are unaware of the rhythms and workings of the divine process of creation. At the present time the traditional Mayan calendar plays a key role in providing knowledge about the plan, and its dissemination is crucial for how things end up for humanity.

As the contours of the cosmic plan become clearer, human beings are presented with a choice as to whether to attempt the climb of the cosmic pyramid. Each one of us has to make this choice individually, and it may require of us that we turn many of our most cherished notions upside down. The kind of choices that humanity makes, collectively speaking, will determine the outcome of creation on this planet. The outcome is not a given. Even if the divine plan is completely predetermined at the level of consciousness, it is not predetermined to what extent human beings will choose to align their lives with its intended end result. In this we really do have a free will. It is also a choice with temptations that look somewhat different for Westerners and Easterners. For Westerners it is a choice whether to seek to maintain the dominance of the world inherited from lower Underworlds or to seek unity, and for Easterners it is a choice of seeking confrontation with the West or to seek unity. For those under the World Tree the choice may be a little easier, but there too the temptation will be strong to settle for the comfort of the Planetary Underworld, rather than to make the final climb to the enlightened state.

Of course, the choice whether to climb the cosmic pyramid is not a one-time choice. Rather it is a choice we make little by little as the cosmic plan unfolds. The choice between unity and duality is yet to be presented to us in many forms and at many times, and as an aid to clarity in making these choices, the Mayan calendar will serve as a very valuable map of the waters we need to navigate. On a deeper level, however, it seems that we will have to make the final choice of path in Day Four of the Galactic Underworld. This is the Day when the effects of the pulsations of the World Tree become more evident and so more starkly present us with the choice of duality versus unity. Certainly, it will be more and more difficult to begin the path toward enlightenment the further the Galactic Underworld has progressed, but in the end, it will be impossible to avoid making the choice.

In this process of making choices, there is also an interaction between our individual paths and the collective path. Because the cosmos can only evolve step by step, the same is also true for ourselves, and so there are things that might be called "time locks." Time locks conceal aspects of our paths until a new calendrical energy allows them to be expressed and creates a new element in our life's path. If we did not have such time locks, we might follow the divine impulses too quickly and only run into a wall. There are crucial points in time defined by the Mayan calendar when portals to a wider consciousness may open for a given individual. As much as we may learn from the Mayan calendar, some aspects of our paths are still likely to remain hidden from most of us—locked up in a secret compartment—until new energies bring them out into the open.

This is a kind of Judgment Day scenario, but in some ways it is different from the scenarios of the historical religions generated by the dualist National Underworld. It does not seem that there will be a separation between the "good" and the "evil" based on how many moral points we have accumulated in our lifetimes, but rather that it is more a matter of developing our intuition to be in the right place at the right time, choosing to surf on the waves of the divine process of creation toward its intended end result of unity, peace, and enlightenment, the New Jerusalem. Those who make such a choice are not necessarily "better" or more valuable than the others. They are just the ones who make this choice.

Yet it may be said that those who do choose to make the climb toward the enlightened state are the more loving individuals, whereas those who choose to maintain duality and dominance, with their many adverse effects, may be too sealed off from the cosmic energies to be able to align with them. Those who decide to climb the pyramid can be expected to want to include as many others as possible simply out of love. Whether this world will end up in a paradise or a catastrophe will not depend on the design of the cosmic plan but on how many choose to make the climb.

The role of free will is thus both greater and smaller than previously thought. On the one hand, we have no say regarding the existence of the cosmic energies of time in the first place. Thus we are not free to create the collective destiny of humanity, which is instead charted by the cosmic plan. This, incidentally, also means there is no such thing as the so-called Hundredth Monkey Effect. Human beings do not create morphogenetic fields—the World Tree and the World Mountain do! On the other hand, we do seem to have free will in deciding whether we want to surf on the waves of creation. There is little to indicate that our choice in this regard is predetermined or that there are any "chosen ones." Nor would it matter what religion we believe in.

Yet there will be differences, within limits, between how rapidly individuals climb the pyramid. Some will reach the top before others. Because the whole cosmos needs to be prepared stepwise and made ready in accordance with the cosmic plan, we must also expect ups and downs until the collective process is complete. There is nothing we can do to change this time schedule and have the whole universe arrive at a paradisiacal state today; at least as a collective, we are subordinated to it. Also, since the plan is perfect as it is, all its steps require time, and none of them can be jumped over. But even if we choose the cosmic plan and enter the cosmic time flow through the proper use of the Mayan calendar, this is not by itself a guarantee that we will attain the enlightened state. For this to happen, we need to develop individual strategies for attaining it.

9
....

The Completion of the Cosmic Plan

ENLIGHTENMENT

Enlightenment is usually associated with Eastern traditions and especially with Buddhism, whose founder, Siddhartha Gautama, attained the enlightened state sometime around the midpoint of the Mayan Great Cycle. His year of birth is variously given as 563, 552, and 534 B.C.E., and it thus seems obvious that his mission had something to do with the ultimate purpose of this cycle, whose end we are now rapidly approaching. Even if very few, if any, of the Buddha's disciples attained the enlightened state, the philosophy he developed, the Middle Way, has played a very important role in Eastern philosophy. After becoming a major philosophy in both India and China, however, it lost most of its influence in the first millennium after Christ.

Figure 9.1. Maitreya, the future Buddha, attaining the enlightened
state after having completed the 108 = (9 x 12) transformations of Shiva,
the Hindu god of creation and destruction

The aspect of Buddhism that may have been the most important for the future of humanity is that it planted an important seed, the idea of attaining the enlightened state, which in much of Eastern thought is seen as the ultimate purpose of life. From our current perspective on the Mayan calendar, it is also not surprising that the influence of Buddhism shrank and that few followers actually attained their spiritual goal, for in the East, as in the West, the National Underworld favored the emergence of dualist religions and philosophies. Based on what we now know about the cosmic plan, we must consider that the time was not yet ready for people to attain the state at which this philosophy aimed.

Today the focus on enlightenment is not limited to Buddhists; it is shared by a much broader group of people, and it does not aim at personal development, healing, or spiritual evolution. Enlightenment is about putting an end to all spiritual seeking, so processes designed to achieve this goal are of a different nature. What makes enlightenment different from the various healing effects of the Galactic Underworld discussed in the previous chapter is that it produces a stable state of joy from which there is no desire to evolve further. In this state there is no desire to become something else, and enlightened people that I have met have been totally present in the moment, radiating joy. They have a highly developed cosmic wisdom and direct contact with the Divine, as well as an apparent freedom from inner conflicts. As they have liberated themselves from the dominance of the past, they do not seem to have any sensitive points that may trigger ego-based emotional reactions. Far from the stereotype popular in the West, they do not sit on Himalayan mountaintops meditating their lives away but are keenly interested in promoting the happiness and enlightenment of humankind.

Although there are already some enlightened people in Europe and the West, it is only natural that the attainment of this state at the present time will occur principally in the East. People of the East are obviously more in resonance with the light of the Eastern Hemisphere. As this light, through the effects of the Galactic Underworld, is added to that of the already dominating Western Hemisphere, the first large groups of enlightened individuals will emerge there. India, the carrier of some of the world's oldest unbroken spiritual traditions, is likely to play a critical role in the enlightenment of the whole planet. It was largely through adhering to these traditions that Gandhi accomplished his monumental mission of helping India attain independence through peaceful means. Because of its distant location in relation to the trunk of the World Tree, India is also one of the parts of the world where the grip of the modern mind is the weakest.

Enlightenment is but one of the levels of consciousness of the cosmic pyramid, the ninth level, and, as always, it is a level some will reach before others. This view of enlightenment as having a history may be seen by some as heretical. There is, however, a tremendous power in the realization that the mind has a history—in fact, one that can be precisely charted as an aspect of the divine plan. It then becomes easily understandable why so many of humankind's earlier ideas for creating peace or better societies have failed. Many ideas have failed to materialize because they did not address the issue of the mind, and it has previously hardly been possible to work toward the wholeness of the mind because the end of creation has been such a long way in the future.

Based on our knowledge of the Mayan calendar, there is little doubt that as we approach the completion of the cosmic plan, the number of enlightened people will increase greatly. In this sense the proof from the Mayan calendar that the mind has a history offers great hope to humankind. When Eastern techniques for transcending the modern mind had been introduced in the West, either in their original forms or in the modern Western varieties mentioned earlier, they presented an essentially static view of the mind. The nature of the mind has been taken as a given, as the way human beings were designed. It is now becoming evident that this view is not accurate, and that human consciousness undergoes an evolution defined by a cosmic time plan. In all considerations about the future of humanity, this time plan now must be taken into account.

There is hope not only because it is becoming realistic for people who choose such a path to attain the enlightened state, but also because if all the problems that plague humanity—violence, wars, stress, low self-esteem, environmental degradation, inequalities, hierarchies of power and money both locally and globally, as well as between men and women, adults and children, human beings and animals, and so on—are created by the dualist mind, then it is becoming realistic for human beings to expect the arrival of a Golden Solar Age based on a transformation of this mind. The separation from the divine source that the dualist frame of consciousness has caused will disappear as a result of the evolution of consciousness, and it has been said that this separation is the sole cause of human suffering. Yet this is hardly something we can expect to happen automatically or as a result of wishful thinking; *the human mind is much more powerful than its thoughts.* The divine plan will require us to manifest concretely in our lives the path of wholeness that the Galactic Underworld makes possible. Needless to say, an intellectual understanding of creation such as is presented in this book is not the same as enlightenment. But it does serve to pave the way for it.

Certainly, there may be different ways to prepare oneself for attaining the enlightened state, but probably not as many as some would like to think. True, we may have different paths of action and individual tasks, so it might seem as if there are any number of paths. Yet if we all share the same mind, it is this same mind that needs to be transformed in all of us, and there is not an infinite number of methods by which this can be done. The mind will not want to disappear of its own accord, so it may easily delude us into thinking we are already enlightened.

Around Kalki, an enlightenment avatar in south India, prophecies have emerged that if humanity is to survive, it needs to attain the enlightened state by the year 2012. Thus, as we are entering the Galactic Underworld, a mission has been set that is expanding to other countries. This mission, I believe, should be seen in the context of the peace mission of Gandhi and the independence of India at the beginning of the Fourth World in 1947. In the larger picture, the purpose of Gandhi's mission in India may have been to prepare for Kalki's mission in the world.

Kalki's teaching, using sutras as a main component, offers a process that is focused directly on enlightenment. Its design is thus different from approaches that merely aim to temporarily transcend or short-circuit the mind temporarily. Unlike any other philosophy of the East, his teaching is grounded in the reality of the cosmic time plan in that there is a set deadline for accomplishing the mission in the year 2012. Becoming enlightened is no longer something the individual seeks out of some desire for his or her own good. Instead it is sought as a means of fulfilling the divine plan and for the good of the rest of humankind. As I see it, this is about to become a necessity rather than a whim.

From the perspective of the Mayan calendar, certain aspects of this development in the East are remarkable. To begin with, Kalki himself was born on a day 13 Ahau (March 7, 1949), which in the Mayan calendar is a much-prophesied energy of enlightenment. Second, Kalki's teaching about the deadline for the enlightenment of humanity by the year 2012 is an independently arising confirmation of the validity of the

Mayan calendar's deadline for the completion of the cosmic plan. The most advanced ancient spiritual traditions of the West and the East, the Mayan and the Vedic, have been unified in a common framework for understanding the future of humanity. The Holy Universe of Time of the West is unified with the timeless Self of the East. What is more, this unification is backed by basic facts of modern science and historical research such as have been presented in this book. Even in the absence of empirical evidence, the convergence of the two traditions would be remarkable and would provide strong support for the Mayan calendar as prophetic. This should be taken seriously by every member of the human species. Even if it does not generate headlines in the media, it is truly history in the making. A rift in the collective consciousness between East and West that has existed for some five thousand years is beginning to be healed.

It is not an accident that in this unification of the ancient traditions, the West contributes the knowledge about time and the East provides the timeless wisdom. There is a difference between East and West in their relationships to time that is also reflected in the functions of the corresponding brain hemispheres. The left brain sequentially organizes the events in life according to a linear cause-effect matrix. Among the ancient traditions of the world it is also unquestionably in the West, among the Maya and the Mexica, that we see the greatest emphasis on time and the most advanced calendars. Calendars in the East were never very advanced, and in China they were linked to such worldly phenomena as imperial dynasties. The right brain mediates an experience of "lived" time, which does not have a clear direction. In India there were long-term calendars, corresponding to the Yuga periods, but these were not exact descriptions of the cosmic plan. In the East the spiritual focus has instead been on the timeless Self and stillness, on just being. That is why the end of creation is often referred to as the end of time. As the dominance of the left-brain hemisphere comes to an end, so will the experience of linear time. Only the present moment will be at hand, and in the enlightened state there will be stillness in the midst of life.

The Mayan calendar, an analytical tool generated by the Western Hemisphere, remains our most important instrument for studying the cosmic plan. We must go this far back in time to find such a tool because modern Western science is a product of the blindfolded consciousness of the Planetary Underworld. This has shut out any spiritual aspect from the scientific study of the evolution of the cosmos, and thus science has been kept in a fragmented state. Only as we enter the Galactic Underworld is the vision of many changing to a degree that they are using the Mayan calendar to study the course of events generated by the cosmic plan. This may help us enter the cosmic time flow leading toward enlightenment.

Even though Eastern civilizations did not develop such an advanced nonphysical calendar system, they have long been aware that the cosmos is subject to a plan. A direct connection to the Mayan calendar system is provided by the holy number 108 that in the East symbolizes wholeness and completeness. This has played a predominant role in Hindu, Buddhist, and Chinese traditions, where it has countless applications. In the Buddhist tradition a rosary, for instance, has 108 beads (ideally made from seeds from the bodhi tree under which Buddha attained enlightenment), and it is said that he who 108 times circumambulates the Holy Mount Kailas, the center of the universe, will reach the enlightened state of nirvana. In China there are 108 movements in t'ai chi and several martial arts traditions, and the Hindus have 108 different names for many of their gods. Creation has supposedly passed through the 108 transformations of Shiva, the god of creation and destruction. What this seems to refer to is that divine creation passes Nine Underworlds, each with twelve transformations between Heavens, which amounts to a total of $9 \times 12 = 108$, the total number of transformations between heavenly energies in the Mayan calendar system. Through this number Easterners expressed the idea that everything was part of the same cosmic plan developed through 108 transformations or movements, 108 energies of time, although they did not develop a calendar to describe precisely the time periods these energies ruled. Regardless, the holiness of the number 108 is a significant link between the spiritual traditions of the East and the Mayan calendar system.

With the Western perspective of the Maya, it is also possible to give a precise meaning to enlightenment: Enlightenment is the consciousness corresponding to the energy of 13 Ahau in all the various Underworlds combined. From the tzolkin charts that have been presented, we can see that 13 Ahau is the energy at which there are no filters blocking the passage of cosmic Light. And so, as all the Underworlds are reaching the tzolkin energy of 13 Ahau, human beings will be in resonance with this enlightening energy. If, as linear time comes to an end, we are able to manifest this, it is tantamount to attaining the enlightened state. On our paths toward this state, with every new Underworld, every new Heaven, and even every new energy of the tzolkin, the possibility for human beings to be channels and expressions of the divine will increase. The passage through the Thirteen Heavens of each Underworld means that one by one, filters and barriers to the passage of divine light will be removed.

The state of enlightenment after the 108 transformations will be different from what it might have appeared to be at any earlier point. This is partly because soon everyone will share it, and also because no reversal will be possible once the divine process of creation is completed. There will be no more lessons to learn and no desire to have an existence dominated by dualities or the fears that they create. There will be no inner conflicts projected onto others, and, as a consequence, there will be both inner and outer harmony and peace. Unlike at the time of Buddha, it will be possible, with proper intention, for everyone to attain the enlightened state, because at this time the cosmos favors such a state. Enlightenment will no longer happen more or less by accident to the lucky few. It is a path that can be taken intentionally; many are already approaching this combined state of happiness and stability.

At our present stage of evolution there are only two more Underworlds to go. The first, the Galactic, essentially serves for preparation and unification; the second, the Universal, is about actually attaining the enlightened state. With every new level of the nine-story cosmic pyramid, we are more truly being created, and actually increasingly creating ourselves, in the image of the Creator. This newfound concordance between East and West provides more than just truth; it also provides hope. If the modern West and the modern East have problems coexisting, it is at least possible to go back to their ancient spiritual origins to find a unified view. Then modern people can take this view as a point of departure from which true equality and global unity can be built anew.

RECOVERING THE LIVING UNIVERSE

As the Galactic Underworld progresses, human beings will increasingly become aware of spiritual energies that have been hidden from them by the blindfolded consciousness of the Planetary Underworld. The discoveries of these energies are already contributing to the emergence of a new worldview in which the universe is not seen as made up merely of dead things. Spirit is about to be recognized as primary to matter, and yet inseparably connected with it. In the emerging worldview the universe is seen as a web of creative, interconnected energies of time and space on different hierarchical levels.

Modern people often frown upon ancient stories about various deities or spiritual entities playing roles in human life. But what we see happening today is that the various energies of Quetzalcoatl, Tezcatlipoca, and so forth are returning. These energies really do interact with the human beings, similarly to what is described in the ancient legends, tricking us into doing certain things and not always being kind to all. If human beings are puppets to these energies, perhaps there is some truth in the ancient legends and the rituals that were developed to communicate with these deities.

With our study it has become evident that there is more than immediately meets the eye to the energies of time. We can empirically verify alternating energies of time in different Underworlds if only we use the Mayan calendar as a template. The reason that people are now increasingly becoming sensitive to the energies of the Mayan calendar is simply that we have entered the Galactic Underworld, where our eyes can no longer be shut to the Divine as the driving factor in the evolution of the cosmos. To the extent that we want to surf on these waves of this cosmic evolution, we will also have to replace calendars based on physical reality with the Mayan.

There is a parallel retrieval of spirituality in our relationships to the energies of space, and there too the Mayan calendar has been an aid for identifying the location of the World Tree, which provides the most important energy grid of the planet. In the same way as there are many hierarchical levels and intricacies connected to the spiritual energies of time, there are many levels of the related spiritual energies of space. Thus there exist a number of different kinds of energy lines, the most notable of which may be the so-called Curry and Hartmann lines that were identified by dowsers in the early 1950s (as the Fourth World created an initial sensitivity to them). The interest in such energy lines, which are not electromagnetic in nature, has, in parallel with interest in the Mayan calendar, increased considerably in recent years.

It seems that the Hartmann lines, running from east to west and from south to north, are microresonances of the World Tree; and the Curry lines, at a 45° angle to them, may be part of the same system. When the Planetary Underworld began in the mid-eighteenth century, we can see from the way houses were built that people then lost their sensitivity to these lines. The Curry and Hartmann lines thus seem to be related to the Mayan Underworlds. Also, it has been found that ancient cliff paintings of Sun Wheels in Sweden, the so-called Crosses of Vuotan (named for the one-eyed father god of the ancient Norse; see fig. 3.12, page 49)—which, similarly to Native American Medicine Wheels, are really representations of the World Tree—are located exactly at Hartmann crosses, indicating that these crosses were regarded as microresonances of the World Tree. How these energy lines change over time in relation to the energy changes of the Mayan calendar is still an important area of study, and it is obvious that at the current time the sensitivity of people is increasing.

If indeed Curry and Hartmann lines are generated by the World Tree, it is not surprising that they cannot be identified as electromagnetic, because the creation fields generated by the World Tree are primary to any physical fields. It also becomes understandable that no one escapes being influenced by the yin/yang polarities of the World Tree, as its field covers the entire planet and is really a field of divine creation.

The World Tree, the central creative principle in the view of the Maya, exists at several different levels of the cosmos, and those that have been discussed in this book, the global and human brains, are only two of them. These are both microcosms of a universal, a galactic, and a solar World Tree. For all we know, all the universe is unified by an invisible web of energy lines related to the creation fields of the cosmos. It is through the existence of this web of lines that changes in energies take place simultaneously throughout the cosmos at the shift points described by the Mayan calendar. We are beginning to recover the view of a pulsating cosmos that is alive and unified through the energies of divine creation. Everything is related, and spirit and matter can no longer be separated.

The World Tree also has a number of holographic microprojections on the human body, where it forms the basis of the chakra system and the various meridians of the Eastern systems of medicine. The seven different chakras can be regarded as a prismatic expansion of the World Tree projection on the level of the human brain, the crown chakra. Several energy systems of the body are generated by these crucial points of energy, forming what may be called a light body. Through the World Tree this energy of life, called *qi* or *prana* depending on the tradition, is related to the creative fields of the entire universe. As the new yin/yang polarity of the Galactic Underworld is beginning to dominate our frame of consciousness, increasing our intuitive sensitivity to its energies, we become able to see the spiritual aspects of our own being. The influence of Eastern systems of healing, and holistic medicine in general, has increased dramatically. Soon it will no longer be possible to separate this from medicine at large. As the Galactic Underworld brings the Cartesian split of body

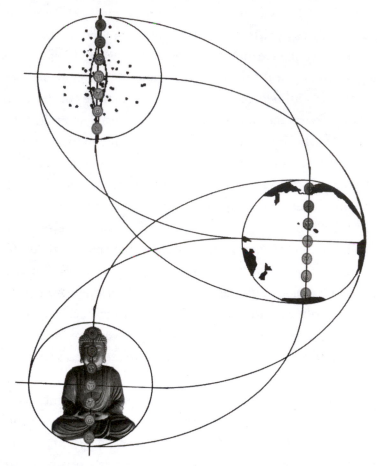

Figure 9.2. Prismatic expansion of the holographic projection of the World Tree into the chakra system, generated by the galaxy and relayed by earth

and soul to an end, the split between mainstream and alternative medicine will meet the same fate.

If our crown chakra is identical to our individual World Tree, it follows that the light polarities and frequencies of this chakra, and its whole prismatic expansion in the chakra system, will vary between different time cycles. The energies of the chakra system will to some extent evolve in synchrony with the Mayan calendar. Typically, as we are now entering the Eighth Underworld, some think an eighth chakra is emerging—an evolution of our light body, or spiritual energy body, in the

Holy Universe of Time. We can thus understand that the ruling tzolkin energy of our day of birth, the day on which we choose to incarnate, has set an energy imprint on us individually that has served to create a unique light body shaping our destiny. This aspect of our light body accounts for the personality traits that are linked to our individual day signs (see appendixes B and C).

The holographic relationships of energy systems on different hierarchical levels of the cosmos seem to indicate that the human being may be too limited a unit when it comes to studying the energy fields affecting human health. Human beings are always part of a larger whole of energies. This has already been proposed by dowsers, who tend to associate certain adverse conditions in the human being with certain types of earth energies. The larger energy fields have been studied systematically in the East; again, with the pulses of yin/yang dualities now favoring the right brain hemisphere, the existence of these spiritual energy fields is becoming more apparent to people everywhere. Ancient Eastern systems such as feng shui in China and Vastu in India are increasingly being practiced in the West as means of harmonizing our surroundings with our spiritual energy fields.

If most of these fields are microresonances of the World Tree, and the World Tree generates the various yin/yang polarities whose coming and going is described by the Mayan calendar, human diseases and their healing may be directly linked to the changing energies of time. This would support the practice of Mayan shamans, who decide the timing of healing practices based on the calendar. The relationship between the healing of diseases and the energies of the Mayan calendar is likely to be explored in more detail as we go deeper into the Galactic Underworld and as our sensitivities to such matters increase. As we progress into the Galactic Underworld, the healing of many diseases may become much easier as the filters blocking the divine light disappear. As we finally come to the energy 13 Ahau in all the Underworlds, divine light will pass through us completely unhindered. This holds the potential for much deeper healing than is currently possible.

But what is true regarding the spiritual nature of the human being is true for planet earth too. The earth itself is dominated by an energy field—a creation field, or light body, that is primary to its material existence. This energy field is fundamentally organized by the global chakra system possibly lined up along the trunk of the World Tree at longitude 12° East, but ley lines, Curry and Hartmann lines, vortexes, nodal points, energy serpents, and so on form meridians of the same spiritual body. Of course, as we will be increasingly aware, this creation field of the planet is nothing but a microcosm of the creation field of the whole universe.

Yet our reception of information from the galactic creation field depends on the purity and balance of the poles of the antennas. The balance of our own brains is very important, but so is the balance of the earth, since this serves as the most immediate relay for the transmission of cosmic information. An ecological perspective, to make the earth into a nonpolluted relay for cosmic information, is thus of paramount importance. In a sense, the current media technology may be blocking the development of intuition and telepathic fields. The greatest threat to the earth and humanity's sustained life on it is the blindfolded consciousness of the Planetary Underworld that looks upon the earth as mere dead matter to be exploited by economic calculations of the left-brain hemisphere. Since this attitude negates the very purpose for which the earth was created, it can only backfire. By a nonpolluted earth, I do not mean only a beautiful, harmonious earth freed of toxins, but also one that is liberated from a number of "harmless" radiations serving for information transmission. Most media originating in the Planetary Underworld, which use electromagnetic waves for dissemination and present a distorted materialist view of the world, are contributing immensely to blocking our ability to receive the true spiritual energies of the cosmos, mediated by the divine creation fields. The media do this primarily through what they focus on and how they present it, but in a sense they also literally pollute the atmosphere with radiation.

The main organizer of the planetary chakra system, the World Tree, pulsates with a rhythm described by the Mayan calendar, and through it the spiritual planetary meridian system and the rhythms of divine creation are directly linked. In a very literal sense, the inner

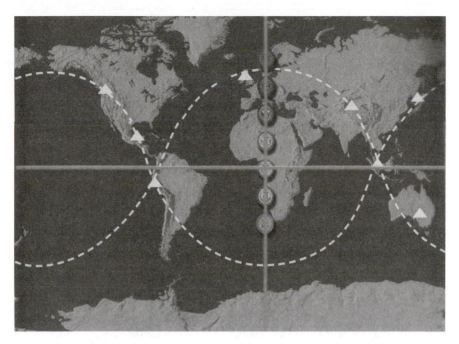

Figure 9.3. Possible global chakra system lined up along the trunk of the World Tree. Also shown is the related double serpent suggested by Robert Coon.

core of Mother Earth has a heartbeat that corresponds to the frequency of change between Heavens. This is the pulse beat of the evolution of our planet charted by the Mayan calendar. The idea of "listening to Mother Earth" takes on a new meaning in this perspective. No longer is Mother Earth only something below us that we graciously take care of. She is also something larger than us with which we need to be in resonance for our evolution toward enlightenment. Our relationship to the earth is not just a matter of "saving the planet" for the sake of its survival. It is a matter of creating an earth of beauty, harmony, and balance that can mediate the energies facilitating our ascension toward cosmic consciousness.

In an ancient prophecy related by Don Alejandro Oxlaj, elder of the Quiché-Maya Council, it is said: "The time of 12 baktun and 13 Ahau is fast approaching, and they shall be here among you to defend Mother Earth." Hence as we reach the tzolkin energy of 13 Ahau, our resonance with Mother Earth will be perfectly developed and our own survival guaranteed.

Why, then, would the energy of 13 Ahau be able to save the earth? Because what is really changing at the current time is our perception of reality, as we are being brought toward a consciousness that will not allow anyone to look at the earth simply as matter to be exploited. This new way of looking at nature and the earth as part of a spiritual web will be irreversibly established with the energy 13 Ahau. A similar change in our vision will take place in our relationship to the sun and the planet Venus. An event is approaching that will be crucial for mediating this very change in apprehension of the cosmos, and the transition to the Mayan calendar.

THE ONENESS CELEBRATION: THE VENUS TRANSITION TO THE MAYAN CALENDAR

The sight of Venus on the Sun is by far the noblest
that astronomy can afford.

EDMUND HALLEY, ASTRONOMER ROYAL, 1691

The sun was always revered by the Maya, as well as by all ancient peoples, who felt a deep kinship with this source of life. The sun has been a symbol and expression of unconditional divine love for as long as humankind has inhabited this planet, always providing it with light and warmth and never asking for anything in return. If you ever doubt that you were meant to be here, this is where to look. The sun has apparently also relayed cosmic and galactic information, conditioning the evolution of consciousness of all things living in the solar system. As we are approach an age of light, the solar age of 13 Ahau, we might wonder what this will mean to our relationship to the sun.

The Maya also very closely studied Venus and its phases and knew, for instance, that five synodical Venus cycles of 584 days equal eight years on earth (minus two days). Venus was seen as a manifestation of Quetzalcoatl, who had once sacrificed himself by throwing himself in the fire and becoming this planet. In various ways the Maya sought to understand how the movement of Venus was linked to the tzolkin. In the Dresden Codex, for instance, the finest of the Mayan codices, the Venus tables play a prominent role. Much of the Mayan interest in Venus was due to the fact that its phases were seen as symbolic of the processes of death and rebirth. It was thought that in the eight days between the disappearance of Venus as the evening star and its emergence as the morning star, Quetzalcoatl would return to the Underworld.

As we are now approaching the enlightened consciousness of the Universal Underworld, our apprehension of reality is changing, and this will include how we view the sun. It is not that the sun, after having been seen as a physical object for centuries, will now come alive as a deity. Rather the consciousness of human beings will change in such a way that everything material will again also be recognized as spiritual. In a sense we will remember our pre-Fall, enlightened consciousness and return to the Garden of Eden. All objects and phenomena will be seen as expressions of the living divine force that pervades the cosmos of which we ourselves are a part. At a much deeper level than is currently the case, the cosmos will be recognized as an expression of the Divine. As all filters that have blocked our vision disappear, as the Universal Underworld draws to a close and all of creation becomes permanently dominated by the energy of 13 Ahau, 13 Light, 13 Lord, or 13 Sun, the cosmos will truly be experienced as alive.

This final transformation of the human mind into a mind of light is heralded by an astronomical event that no person currently alive has previously witnessed, a so-called Venus transit (fig. 9.4) or Venus passage across the surface of the sun. Venus transits mostly occur in pairs, and the last times they could be observed by astronomers were in the years 1761/1769 and 1874/1882. This time around, the initial Venus transit will occur on June 8, 2004 (5.9.2 6 Ik) and the second on June 6, 2012 (2 Ik). Because of modern communications media, most of humanity will be aware of these events. Unlike a solar eclipse, a Venus transit need not be observed from a special location, and in the Eastern Hemisphere and on the East Coast of the United States the first passage

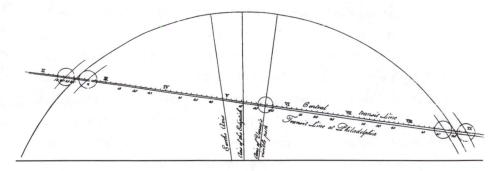

Figure 9.4. The passage of Venus in front of the sun. Drawing by J. Ewing for the American Philosophical Society, as projected in 1769. Such a passage across the disc of the sun takes approximately seven hours.

will be directly visible (through strongly darkened glasses). During this seven-hour-long passage of the planet Venus across the disc of the sun, the sun will serve as a cosmic mirror for humanity. This is an ideal occasion, scheduled by the divine plan, for us to unify in meditation with a focus on the enlightenment of humanity and the transition to the Mayan calendar. It will be a celebration of the spiritual oneness of humanity: the Oneness Celebration.

This event takes place two days after the midpoint of the Third Night of the Galactic Underworld (4 Ahau, June 6, 2004). It heralds the return of the invisible World Tree, the first clear manifestations of which will be evident in its Fourth Day. This Venus transit is an occasion for the collective consciousness of humanity to integrate our fundamental spiritual oneness and the insight that we are now on a path toward enlightenment, where the cosmic plan is designed to lead us. This return to the Underworld by Quetzalcoatl, traditionally the deity credited with inventing the calendar, also means a return of the traditional Mayan calendar, the most exact and advanced expression of the cosmology of ancient Native America. A part of our collective knowledge will be recreated that will serve as guidance to make the global culture whole. People from around the world, regardless of race, gender, religion, or nationality, will participate in this event.

The return of Quetzalcoatl will also mean the return of the wholeness of humankind as the light of the Eastern Hemisphere is added in more tangible ways to the light already shining on the Western Hemisphere. For those seeking an enlightened path, the intention will be to unify the two aspects of our existence at an individual level and integrate the modes of thinking, acting, and being that are mediated by the two hemispheres. Thus a global meditation on this occasion should be focused on unity. There should be focus on integrating the West and the East and transcending the dualities that, both collectively and individually, can divert us from a path toward enlightenment.

This celebration is thus not about astrology, as is usually thought. It is not about expecting that Venus will automatically cause something to happen on earth. Ratherthe cosmos will then offer an opportunity, a unique mirror, in which humanity can see itself reflected as it takes another step on the path toward enlightenment. The Venus transit is an opportunity that means nothing unless it is taken. Its chief role is as a rallying point for oneness, taking place two days after a significant wave shift in the Mayan calendar.

A meditation on the occasion of the Venus transit could take the form of a Gaia meditation. What this means is that you may imagine your own head inside of, or actually becoming, the inner core of the earth. You should then position your eyes at the Hawaiian Islands of this imagined earth, so that your left-brain hemisphere corresponds to the Western Hemisphere and the right-brain hemisphere to the Eastern. In this way you will be fully exposed to the global creation field organized by the World Tree and may in a meditative state develop a resonance with it. You could take this one step further by visualizing your head surrounded by the galactic sphere, with the midplane separating your two brain hemispheres.

The most powerful way of breaking the cultural trance of the materialist calendars is to mark important energy shifts in the Mayan calendar by powerful ceremonies or globally synchronized meditations. Their rhythm should be the same as that once followed by the Mayan ahauob (see fig. 1.5, page 8). They should be performed on tun shifts and half-tun shifts (fig. 9.5), on Ahau days in the true tzolkin calendar. If such a rhythm can be established, beginning on the day 4 Ahau—June 6, 2004—and continue every 180 days, this would contribute immensely to humanity attaining the enlightened state. A breakthrough on a global scale of the rhythm of the Mayan calendar will by itself serve as guidance into the future. It goes without saying that participation in this meditation does not imply adherence to any particular religion, philosophy, or spiritual tradition—nor may any of these be excluded. Through the Venus transits, the cosmos is issuing an invitation to everyone on the planet to partici-

Date	Ruling energy	Tun Count	Phase
June 6, 2004	4 Ahau	5.9.0	Midpoint of Third Night
Dec. 3, 2004	2 Ahau	6.0.0	Beginning of Fourth Day
June 1, 2005	13 Ahau	6.9.0	Midpoint of Fourth Day
Nov. 28, 2005	11 Ahau	7.0.0	Beginning of Fourth Night
May 27, 2006	9 Ahau	7.9.0	Midpoint of Fourth Night
Nov. 23, 2006	7 Ahau	8.0.0	Beginning of Fifth Day
May 22, 2007	5 Ahau	8.9.0	Midpoint of Fifth Day
Nov. 18, 2007	3 Ahau	9.0.0	Beginning of Fifth Night
May 16, 2008	1 Ahau	9.9.0	Midpoint of Fifth Night
Nov. 12, 2008	12 Ahau	10.0.0	Beginning of Sixth Day
May 11, 2009	10 Ahau	10.9.0	Midpoint of Sixth Day
Nov. 7, 2009	8 Ahau	11.0.0	Beginning of Sixth Night
May 6, 2010	6 Ahau	11.9.0	Midpoint of Sixth Night
Nov. 2, 2010	4 Ahau	12.0.0	Beginning of Seventh Day
May 1, 2011	2 Ahau	12.9.0	Midpoint of Seventh Day
Oct. 28, 2011	13 Ahau	13.0.0	Completion of Seventh Day

Figure 9.5. The Venus transition to the Mayan calendar. Shown are Gregorian dates for the Ahau days preceding new half-tuns and tuns of the Galactic Underworld. These Ahau days are very significant opportunities for meditations leading to global synchronization with the divine plan toward enlightenment.

pate in the creation of a collective mind of unity and so prepare us for the attainment of the timeless cosmic consciousness.

Because it is a calendar of the spiritual cosmic flow, the Mayan calendar is fundamentally one of resonance with the divine plan. For those seeking wholeness, the calendar shift, the Venus transition to the Mayan calendar, is about intuitively accessing the energy changes inherent in the cosmic time plan. Organizing and participating in such meditations will contribute to the emergence of a field of intuition and telepathy among human beings. Starting with the day 4 Ahau (June 6, 2004), setting the stage for the celebration of the Venus transit two days later (6 Ik, dissemination), global meditations may be organized on a tun basis according to the time plan above. This will support recognizing the flow of the cosmic time plan with the common focus on the enlightenment of the earth on the day 13 Ahau, October 28, 2011. The transition to the Mayan calendar and the tun-based process of divine creation takes the form of participating in a series of meditations and so creating an alignment with the major energy changes of the Galactic Underworld.

The days June 6–8, 2004, will mark the beginning of the Venus transition to the Mayan calendar, which is the paramount significance of the Venus transits. The Gregorian date June 6 marks a very important half-tun shift on the day 4 Ahau, an energy that derives some of its power from the fact that it was this very energy that brought forth the teachings of Jesus (see fig. 4.6, page 81). Because of its power this tzolkin energy is also what in the Mayan tradition is called a Burner Day, which is still celebrated in Guatemala. This is also the ending energy of the traditional Long Count. The energy of 4 Ahau will be spread even more strongly across the world on the day of dissemination, 6 Ik (6 Ehecatl, the inspiring wind god aspect of Quetzalcoatl), the day of the Venus transit. The activities at the energies 4 Ahau, 5 Imix, and 6 Ik should create in every one of us a consciousness in which both the male and the female aspects, as well as the Eastern and Western aspects, are integrated.

The transition to the Mayan calendar could not come about as the result of a decree, such as the papal decree that instituted the Gregorian calendar in 1582. The energy shifts have to be experienced. Help is needed by broader groups of people to recognize the energy changes that people already subconsciously experience as part of the unfolding of the cosmic plan, and this is what the Oneness Celebration can provide. Only a minority of people is likely to recognize the wave movements of the divine plan and the details of calendrics like those that have been provided in this book. The transition to the Mayan calendar is not, however, a project that can be seen as separate from creating humanity's path toward enlightenment. It is some of the most important guidance humanity will receive regarding its future. *The main rationale for the use of the classical Mayan calendar is that it points out the path toward enlightenment.* To follow, both individually and as a species, the divine tun-based calendar system, as once did the classical Maya, is a means of focusing on our destiny of enlightenment, and as a by-product of enlightenment, on peace. Inasmuch as the Mayan calendar points to the path toward enlightenment, it is also a calendar of peace.

We should not forget that the first Venus transit looks forward to the second Venus transit on June 6, 2012, which occurs at a point in "time" after the cosmic plan has been completed. Even if this date lacks meaning in the Mayan calendar, the second Venus transit may be held up as a beacon for the enlightenment of humanity. In 2012 the Venus transit may serve as an occasion for expressing gratitude to the sun and the divine plan and so serve to stabilize the enlightened state. We may then collectively come to understand the meaning of a cosmic mirror.

THE UNIVERSAL UNDERWORLD

Then I saw a great white throne and Him who was seated on it. . . .
REVELATION 20:11

Again, what will make you realize what the day of Judgment is?
The day on which no soul shall control anything for (another)
soul and the control shall be entirely Allah's.
CLEAVING ASUNDER SURAH 82:17–19, THE QUR'AN

The spiritual beings will remain to create one world
and one nation under one power, that of the Creator.

HOPI PROPHECY

The Universal Underworld in 2011 is what all of creation has been waiting for: its very purpose. It is when all things are brought together and all the conflicting ways of being, acting, and thinking will be resolved and unified in a light that makes it possible for everyone to understand everyone else and everything at once. All limiting thoughts will disappear. The Ninth Underworld may be seen as a gift from God, since it is not only about creating balance, but also the enlightenment given to humankind as an expression of divine grace. This is when we will fully understand why the cosmic plan was designed the way it was and we will overflow with gratitude to the Creator. At the same time, we will recognize our own divinity, for there will be no separation between the divinity of the Creator and our own.

Figure 9.6. We are all one. The universe, our galaxy, our planet,
and all living manifestations are all under the same law of creation and are
synchronized to the energy changes described by the tzolkin. Our destiny
will not be fully experienced before Day Seven of the Universal Underworld.
The last uinal of creation starts October 8 and ends on the tzolkin day 13 Ahau,
October 28, 2011. This is when the purpose of creation is fulfilled.

Not surprisingly, there are several myths from different parts of the world that talk about Nine Worlds. In the Norse tradition, the cosmos was believed to be made up by Nine Worlds, and to the Hopi there are also Nine Worlds. In the unification of science and myth that is now taking place, these and a number of others can be recognized as referring to the Nine Underworlds of the Mesoamerican tradition. What makes the Mayan calendar tradition unique is only its precision.

Paradoxically, as the Mayan calendar system draws to a close, on one hand time will be experienced as moving faster than ever before, and on the other hand as if it is not moving at all. The enlightened consciousness developed in the Universal Underworld will be pulsed onto humankind in a wave movement of the Thirteen Heavens that covers a period of only 260 days (or possibly, if it is one-twentieth of the Galactic Underworld, 234 days; I cannot tell yet). This reflects a frequency of change of the heavenly energies that by far surpasses anything anyone has ever experienced. Yet there is every reason to believe that, at the same point, "time" will come to an end, since "time" is an experience that is predominantly mediated by the left-brain hemisphere. In actuality, the experience of time exists only in a world dominated by the imbalance created by duality. As balance between the two hemispheres is created, instead of time we may expect to experience pure being moment by moment.

This paradox, of course, is not easily resolved by someone whose mind is still dominated by the dualities generated by lower Underworlds. The ego that was established to serve the dualist mind in navigating their changing waters and to maintain a sense of continuity for the individual in the midst of them simply cannot survive at the high frequency of the Universal Underworld. The ego is not consistent with the unitary field of light that will then be ruling the world. In a dualist frame of consciousness, the ego was an important tool for survival; in a unitary frame of consciousness, it will jeopardize the survival of the individual. In an Underworld dominated by unity and enlightenment, acting and thinking based on a dualist frame of mind will become impossible. At the frequency of change that will prevail in the Universal

Underworld, with shifts between Days and Nights taking place every twenty kin, the mind has to be disengaged if it is not to lead the individual to complete heartbreak or personal collapse. Even with proper preparation, for a person who is well aware of the workings of the cosmic plan, the conflict between the five-thousand-year-old dualist mind and the unitary consciousness of the Universal Underworld can only be successfully resolved in one way—through the excision or "slaying" of the ego and the ensuing creation of space for the enlightened way of being to take over. This is what Kalki proposes to help us do. Such external help may be sought and welcomed by some, but others may come to attain the enlightened state simply by surrendering to divine grace. The intent and humility of each one of us will determine the outcome.

As we will then be endowed not only with a galactic but also with a nondualist, universal frame of consciousness, these frames will also make it possible for us to see the events of the past from a different perspective. Being fully immersed in the present, without the dualist mind's need for maintaining continuity with the past, the Universal Underworld will allow for true forgiveness. The Book of Revelation refers to this in verse 21:4: "And God shall wipe away all tears from their eyes; and there shall be no more death, neither sorrow, nor crying, neither shall there be any more pain; for the former things are passed away." From the perspective of the Universal Underworld, the former things will have passed away.

By 2011 the dominance of the dualist mind will wither, and all the conflicts of humanity originating in lower levels of consciousness will be resolved. From the perspective of the enlightened state, the old order will no longer be real. The advanced technologies developed by the dualist mind will find the place they were always meant to have, in service to humanity and the living cosmos, not as tools of domination. By then not only the old monarchic rule, but also democracy, will be a thing of the past (if everyone lives in unity and harmony with the Divine, why elect someone to rule them?). All hierarchies will have crumbled. With the end of duality, the dominance of one soul over any other will naturally come to an end, and so there will be no need for a government (which is like the ego of the nation) to steer people through conflicting interests. All human beings will, in a much deeper sense than at present, be recognized as having equal value—each as her or his own manifestation of the Divine. In the process of climbing to the universal level of consciousness, all limiting thoughts will disappear.

Once this has happened, we will be able to fully experience the unity with All That Is moment by moment, or frame by frame, as it were. Such an experience, frame by frame of consciousness, will allow for total sensual enjoyment in the present moment. Lacking the mind's need to maintain continuity, we will experience complete freedom, and our status as puppets of the divine process of creation will come to an end. No longer will there be an experience of separation between human beings and God. If we do not experience ourselves as gods, we will, at the very least, experience ourselves fully as the manifestations of the Divine that we truly are.

As a corollary, to wait around to see what will happen in the year 2012 is to totally miss the point. It will simply not be possible not to be enlightened after October 28, 2011, or at least not from a certain time afterward when the new reality has definitely manifested. With a dualist mind, it will not be possible to be in resonance with the new unitary divine reality under one power, the Creator. It would thus seem wise for all of us to prepare ourselves, beginning today, by immersing ourselves in the cosmic flow of time and in all possible ways seeking to transcend the influence of dualist Underworlds on our thinking, acting, and being. After all, the Universal Underworld will favor an enlightened state of being of love and joy, and once this has been established no reversal to duality will be possible.

This will mean the return of an apprehension of the living cosmos. The enlightenment that this cycle brings will allow for true and complete healing and forgiveness of the past, which humans will also extend to God, whose relationship with human beings (because of the duality of lower Underworlds) has sometimes been perceived as antagonistic. When the energy of 13 Ahau is reached in all the Underworlds, no filters will block our vision of divine light or our communion with the

Divine. We will come to live in the New Jerusalem, having reached enlightenment after climbing the Nine Underworlds and completing the 108 movements of Shiva. After the building of its ninth level has been completed, humankind will be ready to stand on top of the cosmic pyramid. Humanity will be complete, living fully in the present at a higher level of awareness and with a pure enjoyment of living. The creation cycles that have served to build the cosmic pyramid, partly by making human beings into puppets, will come to an end. The life of Universal Humankind will begin.

. . . AND BEYOND

As we have seen in the history of humankind, it takes a certain amount of time until a new frame of consciousness finds its full manifestation. It is only realistic to expect that something similar will be the case after the enlightened universal frame of consciousness is finally established as a Heaven on the day 13. 13. 13. 13. 13. 13. 13. 13. 13. 13 Ahau (October 28, 2011). Even if the status of human beings as puppets of creation disappears through the influence of this energy, it seems unlikely that this will immediately be expressed in stable forms. Hence the year 2012 may also be a period during which many people will have to find ways of adapting to the new frame of cosmic consciousness. If nothing else, we will need to adapt to the fact that everyone around is now enlightened and has full faith that a millennium of peace, the Golden Solar Age, has finally dawned on earth. *This does not mean that a "new cycle" will begin.* It is the end of cycles.

With the ego slain in the Universal Underworld, the Heavens will favor a consciousness that is devoid of inner conflicts and produces no external conflicts. The year 2012 in particular, however, will be required for this enlightened state to settle in a number of individuals who may not have been fully prepared for it. For the final settling of the enlightened consciousness of the Universal Underworld, the celebrations on the occasion of the Venus transit of June 6, 2012, will be very important.

They will be expressions of gratitude to the divine cosmos for having brought us to where we are.

Both Mayan and Christian sources talk about an end to death at the end of time, and Eastern traditions talk about the enlightened state as deathless. The ancient traditions all point in the same direction: a timeless, enlightened state of cosmic consciousness. We may then rightly ask, who in such a state needs a calendar? The Mayan calendar will have come to an end, and its use becomes meaningless. The calendar is like a staircase that is absolutely necessary to bring us to the top but serves no purpose after we reach it. In the year 2011, at a point when all Underworlds will have attained the tzolkin energy of 13 Ahau, the divine process of creation is completed. What the "future" will be is not predictable from the Mayan calendar, because the wave movement of all its Underworlds will have come to an end. It is the nature of the enlightened human being to be totally free to live according to his or her choosing. With such a freedom, and liberation from the cosmic plan, the further destiny of humankind is impossible to predict. We will be completely free to chart our own destiny. Humanity will live in true freedom, joy, and peace.

APPENDIX A

The Cycles of Economy

If the Mayan calendar describes the wave movement of the evolution of human consciousness, one would assume that it also has something to say about how the ups and downs of the world economy are generated. That no one has discussed this previously could give rise to a false perception that the Mayan calendar does not describe economic changes.

For lack of a better word, I will refer to the modern economic system as capitalist even though this term is loaded by its proponents and adversaries with conflicting implications. Regardless of what we call it, today's global economic system has undergone a significant transformation as a result of the influence of the creative pulses from the World Tree. It is not a static system that can be placed in a box. But as of January 5, 1999, we have entered a new Underworld, so it is reasonable to expect very significant transformations of this system. It is timely to ask what will happen with the abstract "capitalist" world economy.

Some may object to the word *abstract*, maintaining that the world economy has a tangible physical nature, as goods are being transported and sold all over the world and money is shifting hands. Yet the driving forces of today's economy are abstract. To recognize the fundamentally intangible nature of the capitalist economy, let us engage in a thought experiment: Assume that, for whatever reason, all the information in the computers of the banks of the world was suddenly erased, meaning that all accounts of holdings of money, stock, options, and so on would disappear. Although many would regard this as a catastrophe, would anything of real value be lost? Obviously not! All natural resources, buildings, machines, goods, human knowledge, and so forth that existed before would remain untouched by such a hypothetical synchronistic computer crash. In terms of real values, nothing would be lost, and the world could easily pick up the day after (leaving aside the emotional effects that this occurrence might have).

Of course, if nothing of real value, only abstract values, would be lost, there is only one logical conclusion to be drawn: All the paper and cybermoney that is currently circulating in the world's financial markets is completely worthless. The abstract world economy is really a big Monopoly game, which continues for the sole reason that people still agree on the rules—that paper or cyber money (digits in bank computers) has value. But the agreement on its value is only an intellectual one.

This has consequences. First, since the oscillations of the World Tree described by the Mayan calendar play the predominant role in shaping the mental history of human beings, this also means that economic cycles are directly dependent on the various baktun and katun shifts of this calendar that reflect these oscillations. Second, an economy based merely on mental agreements must be very sensitive to these oscillations that shape the nature of our minds in any given Underworld. Especially as new Underworlds begin, radical changes in how the economy is organized must be expected.

Obviously, this point must be qualified by what we know from the evolution of the world economy in the past. For this we may start by looking at how this has been influenced by the changing baktuns of the Mayan Long Count. Technological development came about as a result of the resonance of the human consciousness with the waves generated by the World Tree. I will not detail this development here. Suffice it to say that around the time of the beginning of the Great Cycle in 3115

B.C.E., bronze was used for the first time in Crete, Sumer, and Anatolia, and so human beings left the Stone Age. In spiritual terms, the beginning of this new National Underworld (developed by the thirteen baktuns of its Great Cycle) meant that the World Tree (Tree of Life, oscillating through longitude 12° East) generated a global dualist creation field that in the Book of Genesis is metaphorically referred to as the Fall. On the positive side, this field, favoring the Western/analytical hemisphere, gave rise to technological advances, including metalworking (metaphorically, this was knowledge gained from eating the fruit of the Tree of the Knowledge of Good and Evil). Metalworking, unlike any earlier craft, was such a specialized activity that trade, at least in the form of barter, became necessary. On the negative side, however, this dualist frame of consciousness generated organized warfare as well as an unequal class society where people were valued differently (after eating of the fruit, Adam and Eve were able to see good and evil, according to the Book of Genesis). Both the creativity and the separation were actually projections of the dualist mind that the World Tree began to generate at the inception of the National Underworld.

What made the world economy into a Monopoly game was not barter, but the evolution of monetary instruments, the means of tender. Figure A.1 shows how the various means of exchange have developed pulsewise during the Days, the seven baktuns dominated by duality. With every pulse favoring the left-brain hemisphere (with every Day), humanity took a step in developing abstract, standardized means of exchanging value. The evolution goes from weighed pieces of silver in the first day to paper bills in the seventh, with the first real gold coins made exactly at the midpoint of the whole Underworld. Thus we can see how the pulses from the World Tree have influenced the human mind to create increasingly more abstract means of exchange.

The Days are the periods when the World Tree favors creativity. From this we may understand how the abstracting economy came into existence in the first place. The final step was taken as the completion of the Thirteenth Heaven (ruled, according to the Aztecs, by the dual god Ometeotl/Omecinatl) began in the early seventeenth century. Sweden, as

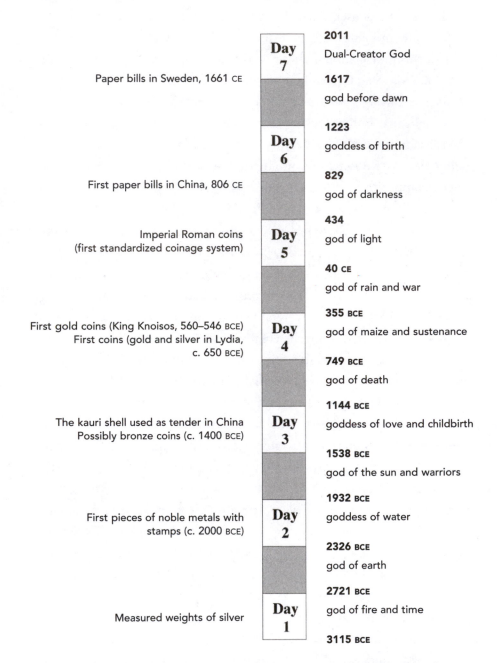

Figure A.1. The National Underworld and its development of tender

the first country in Europe, then implemented the use of paper bills in 1661. China, of course, had used paper money several centuries earlier, but it was somewhat different. The Swedish paper bills were not receipts written in the name of the holders for specific sums, but anonymous notes with different denominations. This initial experiment was somewhat ahead of its time and failed because of the ensuing rapid inflation (the head of the Central Bank was put into jail because of it). The idea of paper bills was taken up again by the Bank of England in 1694 and spread to its American colonies in the eighteenth century.

Generally, the rule by Ometeotl/Omecinatl of this Heaven meant a strong push forward for abstract thinking: the scientific revolution, Protestantism, and paper money. These are mental constructs typical of the left-brain hemisphere, through which analytical reasoning and mathematical calculations are mediated. In parallel with this, and related to it, the rule of Ometeotl/Omecinatl meant the beginning of a four-hundred-year period of world dominance by the West; the buildup of the British Empire began at this time. Because the mind-set ruling this baktun favored both the Western Hemisphere and, through holographic resonance, the left-brain hemisphere, it is no wonder that paper money, after its initial use in Sweden, spread principally to the West. In fact, many other European countries did not begin to use paper bills until the beginning of the nineteenth century.

Mentally speaking, the shift taking place with the use of paper bills was very significant. At this point, the values of life came to be translated into abstract quantities written on pieces of paper. Of course, in this early time there was still a guarantee that the value of the bill could be traded for gold. Nonetheless, the use of paper money reflected one of the significant steps in the alienation of human beings from nature (an aspect of the Fall, which was completed with the Thirteenth Heaven) and the beginning of an abstract, detached view of the earth as matter to be exploited based on the abstract calculations of the left-brain hemisphere. While the use of paper money set the stage for the emergence of capitalism, a second step was necessary for its further development: the Industrial Revolution.

The Industrial Revolution began with the Planetary Underworld in 1755, heralded by the first major steps in industrial development, the spinning jenny (1764) and James Watt's steam engine (1769). The further industrial development that occurred at the beginning of this Underworld's Days is beyond the scope of this appendix, which deals with economics rather than technology.

To study economic cycles we should first note the times of significant economic crises taking place within this Underworld, which are indicated in the timeline in figure A.2. Included in this timeline are only international economic crises—crises that occurred in more than one major country simultaneously. Notice first of all that the Nights in this Underworld invariably begin with recessions or depressions.

This is very evident if we consider the beginnings of its later Nights—1893, 1932, and 1972—which are the best known among people living today. The depression in 1893 was called the Great Depression in the United States, because people at the time had never seen anything like it. But the stock exchange crash in 1929 and the depression that followed were far worse, so that prolonged crisis took over the "Great Depression" designation. The worst year in international trade of this period was 1932, the year the Fifth Night began. That the worst depression in history would begin with the rule of Tezcatlipoca, the lord of darkness, should not be surprising to those familiar with the different energies of the Mayan calendar. Nineteen seventy-two was the first year of slower growth after the uninterrupted economic expansion that followed World War II. This was seemingly initiated by the energy crisis, but in reality it was triggered by the mental influence that the new Night exerted.

In figure A.2 the Days, in contrast, are periods of more or less sustained growth. Strong periods of growth are very obvious in Days Six and Seven. Again, this is not surprising given that the Days are periods when the World Tree favors mental creativity. The Seventh Day began with a very long period of uninterrupted growth in the West and especially the United States. In the 1990s, economic policy seemed very successful, as a number of innovations in the computer industry and

Economic events (left)	Day/Night	Year and deity (right)
	Day 7	**2011** Dual-Creator God
Beginning of period of fast growth (1991)		**1992** god before dawn
Beginning of period of slower growth (energy crisis, 1973–1975)	**Day 6**	**1972** goddess of birth
Beginning of period of accelerated growth (early 1950s)		**1952** god of darkness
The Great Depression (1929 in U.S.; 1931–1933 in Europe)	**Day 5**	**1932** god of light
Economic crisis in France (1907)		**1913** god of rain and war
	Day 4	**1893** god of maize and sustenance
Economic crisis in Germany and the U.S. (1873)		**1873** god of death
Economic crisis in the U.S., England, and continental Europe (1857)	**Day 3**	**1854** goddess of love and childbirth
Economic crisis in England (1836), the U.S. (1837), and France (1838)		**1834** god of the sun and warriors
Economic crisis in England and the U.S. (1819)	**Day 2**	**1814** goddess of water
		1794 god of earth
Economic crisis in London and Amsterdam (1772)	**Day 1**	**1775** god of fire and time
		1755

Figure A.2. The Planetary Underworld and its economic recessions and periods of growth with the ruling deities of the Days and Nights

telecommunications created a mental confidence that growth would go on forever. As we can see in figure A.2, however, some of the Days early in this Underworld began with economic crises because the new products and ideas that emerged as a result of these new pulses of creativity generated a shortage of financial means. So also the beginnings of Days have sometimes seen financial crises, but they were not as difficult as those that occurred at the beginnings of the Nights of this Underworld.

Because of its origin in these pulses, the world economy has developed, at least since the mid-nineteenth century, according to a katun cycle (twenty tun or 19.7 years) for economic growth. A twenty-year cycle for capital accumulation in the United States following this pattern has been demonstrated by Harvard economist Simon Kuznets. As we have progressed into the twentieth century, growth cycles of shorter duration have increasingly been superimposed on longer cycles, most with growth periods of about five and ten years, respectively. These may be seen as overtones of the katun cycles, and as the Galactic Underworld came closer these high-frequency economic cycles, reflections of the Mayan *holtun* (five tuns) and *lahuntun* (ten tuns) periods have become more marked. Until recently, the five-year economic cycle was dominant in the world economy. As the ancients knew, the vibrations of the World Tree are behind all things, and in ancient Scandinavia it was said that as creation approached its completion, its branches would shiver faster—and here we see the results.

We should keep in mind that the way we are looking at economic cycles is different from the view of traditional economic scientists, who have never been able to explain the existence of economic cycles in the first place. The Mayan time cycles are not endlessly repeated; they have a definite timeline—October 28, 2011—at which point the world will have arrived at a balance, and the cyclical changes in the economy will have come to an end. (The "economy" at that time will be in service of the enlightened; the stressful influences of a fluctuating economy will no longer exist.)

Now that we have seen that the basic cyclical changes in the economy are a function of the wave movement described by the Mayan calendar, we can return to the issue of the abstraction of economic values. A key indicator for this abstraction is how much of a link is retained between the global Monopoly game and real values—that is, to what extent paper money can be exchanged for gold (the gold standard). It is interesting to follow major steps in the Planetary Underworld in this regard using the U.S. economy (which, after all, epitomizes Western capitalism) as an example (fig. A.3). The pattern that emerges is that the beginnings of Days mean a strengthening of the link between gold and dollars, whereas the beginnings of Nights mean the link weakens. In the beginning of this Underworld, when dollars were made from noble metals, there could be no abstraction of value. Yet as the gold standard has been abolished at the beginnings of Nights, money has lost its direct link to real values. From 1972 on, the monetary system of the world has not been based on a gold standard, and so it has come to lack a foundation in real values.

Figure A.3, in fact, is probably one of the most striking manifestations of the wave movement of the divine plan that has ever been discovered. It might be objected that the meeting in Bretton Woods, at which the post–World War II global financial system was outlined, took place in 1944, which does not conform to the Mayan calendar. But this meeting only decided on a number of ideas whose implementation did not begin until much later. Even by 1950 none of them had been put into effect; until the beginning of the 1950s, several of the Western European countries still did not have convertible currencies. The Bretton Woods system—according to which the United States guaranteed foreign governments that it would exchange their dollar assets for gold—took effect around 1952, as the Sixth Day of this Underworld began. The Mayan calendar usually does not describe when ideas are conceived, but when they actually manifest. At the end of this Day, in 1972, the gold standard was effectively abolished from the world's financial system.

What should be absolutely clear from these figures is that the pulses of the World Tree directly determine economic development. The correlation between the two is so strong that most economists could only dream of attaining a similar concordance to support a theory. What is more, it is based on very simple facts that anyone with a basic economic

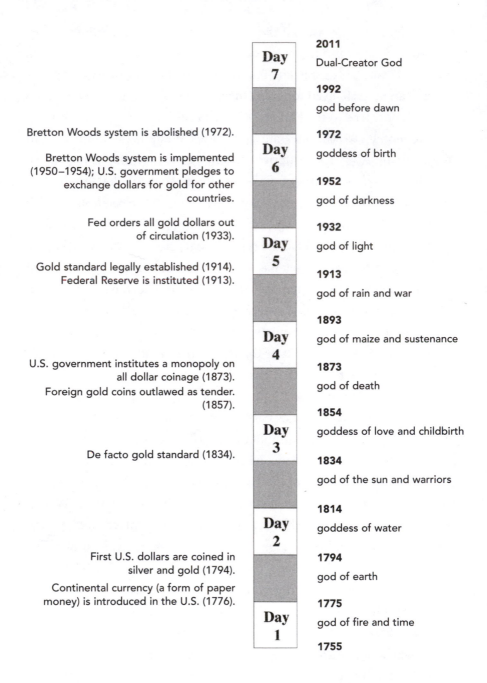

	2011 Dual-Creator God
Day 7	
	1992 god before dawn
Bretton Woods system is abolished (1972).	**1972** goddess of birth
Day 6	
Bretton Woods system is implemented (1950–1954); U.S. government pledges to exchange dollars for gold for other countries.	**1952** god of darkness
Fed orders all gold dollars out of circulation (1933).	**1932** god of light
Day 5	
Gold standard legally established (1914). Federal Reserve is instituted (1913).	**1913** god of rain and war
	1893 god of maize and sustenance
Day 4	
U.S. government institutes a monopoly on all dollar coinage (1873). Foreign gold coins outlawed as tender. (1857).	**1873** god of death
	1854 goddess of love and childbirth
Day 3	
De facto gold standard (1834).	**1834** god of the sun and warriors
	1814 goddess of water
Day 2	
First U.S. dollars are coined in silver and gold (1794).	**1794** god of earth
Continental currency (a form of paper money) is introduced in the U.S. (1776).	**1775** god of fire and time
Day 1	
	1755

Figure A.3. The Planetary Underworld, the changing relationship between the dollar and gold, and the ruling deities of the Days and Nights

education can verify. The pattern leaves us with an obvious exception, however—the beginning of the Seventh Day of the Planetary Underworld in 1992 when no aspect of the gold standard was introduced. Thus the long uninterrupted period of economic growth that followed was not based on any gold standard.

There is a parallel between the Days ruled by Ometeotl/Omecinatl in both the National and Planetary Underworlds. In the Seventh Day of the National Underworld, paper money came into use in the West, which meant an abstraction of real values. When the Seventh Day of the Planetary Underworld began, as a further step this paper money lost all relation to real values, in this case gold. The world economy has thus come to be based on a monetary system that, as I concluded earlier, is totally worthless. (I am half-joking, of course, as I use money myself, but in the big picture this is really true.)

What is behind this pattern? It is well known that the purpose of disconnecting the value of money from gold when economic depressions begin is to keep the wheels of industry turning by artificial means (printing paper money). What is less well known is that these depressions have come about as a result of the mental effects of the Nights in the Mayan calendar. As human industrial creativity has increased at the beginning of new Days, in contrast, there is a risk of economic overheating, and hence on such occasions there has been a tendency to reinstate the gold standard to keep things grounded. Regardless, the effects of abolishing the gold standard for long periods, and a number of mechanisms in banking that I will not go into, are that today's monetary system is a mental construct floating in the air based on the confidence people have in it.

In 1992 the European Union devised a plan to launch the euro currency, a decision that was implemented exactly as the new and higher Galactic Underworld began, in January 1999. In a sense, this is an even farther step away from a link to gold; unlike the United States, the European Union does not hold a common reserve of gold, and the value of the euro is based only on the confidence people have in the combined assets of the member countries. The financial system of today's world has to an extreme degree been disconnected from the value of anything real. This is obvious in the United States, where paper money cannot be exchanged for gold, but it is even more obvious with the European currency, where it is difficult to identify what national assets are used to back its value.

This brings us to the question of how the "economy" will develop in this Galactic Underworld. If the euro was implemented on its First Day (January 4, 1999), one might ask whether this had a greater significance than immediately meets the eye. If, as we have reason to expect, the Days will bring an end to the world dominance by the West (use the timeline in fig. 7.4, page 155, for reference), it is entirely possible that the new currency will be part of the buffering role that Europe is about to take in the conflicts between East and West. It seems likely that the Days will strengthen the euro at the expense of the dollar, and that during Nights, the dollar will regain lost ground.

In terms of technological development, the First Day of the Galactic Underworld (essentially the year 1999) generated the most enthusiasm about information technology and the "new economy," amounting to a revolution on a par with the Industrial Revolution. Given the parallel to the First Day of the Planetary Underworld, it is easy to understand why this was a widespread impulse. The inspiration it gave to investors meant that everything that sounded like *Internet* or *telecom* soared on the stock markets.

The Days of this emerging new Underworld may bring economic ups and the Nights downs, quite in parallel with the pattern we discovered for the two lower Underworlds, the National and the Planetary. So far this pattern has proved generally accurate, and Night Two, essentially the year 2002, has, as expected from its energy, displayed a dramatic decline in all stocks related to telecommunications, punctuated by a series of scandals involving accounting fraud. In the IT industry we have a pattern that, albeit at the higher tun-based frequency of the Galactic Underworld, is a direct parallel to that of the lower Underworlds, where Days have been conducive to economic innovations and Nights to recessions.

Yet maybe we will not be served economic predictions on such a polished silver plate. After all, each Underworld develops a new and distinct frame of consciousness, with specific characteristics, so entirely unexpected phenomena often arise. We must therefore be cautious in our interpretations. For one thing, parallels with lower Underworlds are complicated by the fact that the Galactic Underworld will not favor the left-brain hemisphere. Even if the IT industry is then favored by the Days, we must ask what will happen to the abstract, "worthless" global monetary system, for the pulses of the Days will also favor the right-brain hemisphere. Even if the Days remain periods of technological creativity, the emerging creativity will not be the abstracting creativity of the left-brain hemisphere. And, of course, if the more tangible and concrete ways of looking at reality that are typical of the right-brain hemisphere are now about to be strengthened, perhaps there will be a collapse of the global Monopoly game. This could happen in any number of ways, possibly triggered by the activities of hackers or computer failures. Regardless of the forms that such a collapse may take, it seems that the best bet is for it to occur close to the time that the Fifth Night begins, in November 2007. What is absolutely clear from the foregoing is that the economic cycles will proceed in phase with the Mayan calendar.

Practical Uses of the Mayan Calendar

THE REVIVAL OF THE TZOLKIN

The tzolkin may be used on two different levels. One is the cosmic, or global, level, where it can be used as a basis for prophecy. That is what has been discussed in the main body of this book, for instance by using the seven Days and six Nights that are part of the tzolkin pattern (see fig. 6.1, page 121). The other is the individual level, where its use is akin to divination. On a spiritual level the two uses are directly linked. Unless prophecy is possible, divination would not be possible, and vice versa. Yet in recent centuries it seems that among the Maya the role of prophecy based on the tzolkin has been downplayed. Today the tzolkin has four main functions in its region of origin, which are partially related. One such function is to tell us the nature of the tzolkin energy that rules a given day, to give us an idea of what themes may surface on that day. Knowledge of these themes is often linked to recommendations for things to do on the different days.

Very much related to this is the second function of the tzolkin as a cycle of religious festivals occurring on special

Figure B.1. A shaman woman advising two persons in dispute and performing divination by shell casting

tzolkin days. While the ecclesiastical year of the Christian churches is based on the Gregorian calendar, the spiritual "year" of the Maya is the tzolkin. One of the holy days in the tzolkin is, for instance, 8 Batz (or 8 Chuen), when the Sacred Calendar is celebrated in Momostenango. Aspirants to become day-keepers are then ordained and begin to receive special training about the calendar throughout the following 260 days. The tzolkin round also includes other holidays, when other matters in the Mayan world are celebrated.

The third function of the tzolkin is for divination by shamans. In principle, this is not unlike drawing cards from a tarot deck, except that the symbols of a tzolkin chart are used. In such divination the day signs and numbers have specific meanings, which are partly the secrets of shamans.

The fourth function is to provide information about birthdays and what these mean regarding personality traits and destinies. Because every day has a special energy in the tzolkin count, we can learn about ourselves from the energy in which we incarnated at birth. In what follows I will give only some hints about the meanings of personal tzolkin combinations and indicate how the reader may continue his or her exploration. We should recall that the revival of the true Mayan tzolkin count is very fresh, and outside of Guatemala the Sacred Calendar has hardly been followed anywhere for more than ten years. It is thus still waiting to take its rightful place in the heritage of the planet.

Anyone taking up the use of this calendar is thus a participant in a pioneering exploration. To use the tzolkin means for modern people to reconnect with the process of cosmic creation, the source of our existence. The tzolkin can help us become aware of the changing energies dominating different time periods and allow us to be supported by them.

THE TWENTY DAY SIGNS

If the Mayan calendar is, as we have seen in this book, a calendar of the cosmic creation process, it would mean that, properly used, it is a tool to help us align our individual lives with this process and so with our true purpose in life. In fact, we should ask ourselves if we would even have an individual purpose in life unless we were living in a creation with an overall purpose developed by a cosmic plan. After all, it is the cosmic plan that determines that some things are meant to happen and others not. To experience more of the former types of events, we should use a calendar that describes the flow of divine creation and seek to harmonize our own paths with the flow of the larger whole.

The Thirteen Heavens and the thirteen-day count have been extensively discussed. The tzolkin is also made up of twenty-day cycles, uinals, which are generated by progressions through the twenty glyphs (also called Day Lords or day signs; see fig. B.2). In divinatory work more emphasis is usually placed on someone's day sign of birth than on her/his number, so in the present context the former are discussed.

There is extensive traditional information as to the energy of each of the Day Lords in the twenty-day cycle. The nature of this energy is partly encoded in the glyph that is used to symbolize it. This information usually has the character of specific themes that may dominate the rule of a certain day sign. Although the Maya, Aztec, and Cherokee tzolkin day-counts are synchronized, there are differences in the symbols used for the day signs that are partly related to the different environments in which these different traditions emerged (Jaguar among the Maya is Ocelot among the Mexica and Panther among the Cherokee). Yet the archetypal influences that the day signs reflect are obviously the same. For those seeking a deeper knowledge of the traditional meanings of the day signs, it may be advisable to study them from the different viewpoints provided by the different traditions. Kenneth Johnson's *Jaguar Wisdom* describes some of the Mayan views, Bruce Scofield's *Day Signs* much of the Aztec, and Raven Hail's *The Cherokee Sacred Calendar* the Cherokee view of the day signs.

In these books the day signs are presented with some of the traditional lore of the respective peoples. Certainly, many day keepers and shamans belonging to the different traditions may have a great deal to add in this respect. Day-keepers have inherited and accumulated much knowledge of the magical meanings of day signs.

Similarly to the thirteen numbers, the twenty glyphs are also associated with a number of deities (fig. B.2) that may be seen as contributing to the energies of the respective days. As with the thirteen numbers, valuable information may be gained about the energies of the twenty glyphs through the study of the characters of the deities that rule these glyphs. In this figure are also listed the names of the corresponding totem animals according to today's Quiché-Maya.

In the modern world, by far the most common birthday to celebrate is that based on the seasonal year. Someone born on May 12, 1951, for instance, celebrates his birthday on the twelfth of May every Gregorian year. This way of celebrating birthdays is by no means to be taken as a given. Celebrating one's birthday in the physical year obviously correlates with a view of the human being as primarily a component in a

Mayan Day sign	Aztec Day sign	Ruling Aztec Deity	Nagual (Totem)
Imix	Cipactli	**Tonacatecuhtli,** god of procreation	Fish or alligator
Ik	Ehecatl	**Ehecatl,** god of wind Quetzalcoatl is in this guise	Weasel or lynx
Akbal	Calli	**Tepeyollotl,** the heart of the mountain	Fawn
Kan	Cuetzpallin	**Huehuecoyotl,** god of dance	Lizard
Chicchan	Coatl	**Chalchiuhtlicue,** goddess of water	Serpent
Cimi	Miquiztli	**Tecciztecatl,** goddess of the moon	Owl
Manik	Mazatl	**Tlaloc,** god of rain and war	Deer
Lamat	Tochtli	**Mayahuel,** goddess of pulque	Rabbit
Muluc	Atl	**Xiuhtecuhtli,** god of fire and time	Shark
Oc	Itzcuintli	**Mictlantecuhtli,** god of death	Wild dog

Figure B.2. The twenty Day Lords with their ruling deities and *naguals* (totems)

physical universe, rather than a spirit that incarnates in a Universe of Holy Time.

The Maya celebrated birthdays based on the spiritual energies of their day of birth, which are cyclically repeated every 260 or 360 days. This way of celebrating birthdays implies a totally different view of what it means to be a human being. Celebrating birthdays in spiritually based time, in the tun or the tzolkin cycles, implies that one sees human

Mayan Day sign	Aztec Day sign	Ruling Aztec Deity	Nagual (Totem)
Chuen	Ozomatli	**Xochipilli,** god of flowers	Monkey
Eb	Malinalli	**Patecatl,** god of medicine	Thrush
Ben	Acatl	**Tezcatlipoca,** god of darkness	Bee
Ix	Ocelotl	**Tlacolteotl,** goddess of love and childbirth	Jaguar
Men	Cuauhtli	**Tezcatlipoca, Xipe Totec,** god of darkness, flayed god	Quetzal
Cib	Cozcacuauhtli	**Itzpapalotl,** obsidian butterfly	Vulture
Caban	Ollin	**Xolotl, Nanahuatzin,** dogheaded monster, sun	Woodpecker
Etznab	Tecpatl	**Chalchiuhtotolin,** turkey god and a bird variety of Tezcatlipoca	Wolf
Cauac	Quiahuitl	**Tonatiuh,** god of the sun and warriors	Lion
Ahau	Xochitl	**Xochiquetzal,** goddess of flowers	Eagle

Figure B.2 *(continued)*

beings primarily as spiritual beings. And because it is the energies of the consciousness cycles that make this creation evolve, it may be of greater interest to know when one was born in the context of evolutionary spiritual time than in the context of endlessly repeating astronomical cycles.

Even today the Maya often give a person the name of the combination of the particular tzolkin day she was born. This is because your particular tzolkin birthday may be important for your destiny and who you

Figure B.3. The Aztec deity Tonacatecuhtli with the day-sign
Cipacti (alligator) that he ruled

are as an individual. By the method outlined in appendix C, you may calculate your tzolkin energy according to the classical Quiché-Maya-Aztec-Cherokee count from your day of birth in the Gregorian calendar. It is also possible to download calculators for this purpose from the Internet. (Please use correlation 584, 283 if you are interested in the count still in use by the Maya. Some archeologists, however, still favor the 584, 285 count, which I discourage.) For such calculations there are different ideas about when the day begins. My personal experience, mainly from working in Sweden where the length of day varies more than in most other places, has led me to set the beginning of the day at sunrise.

Some character traits often linked to individuals born on a given day sign are given in figure B.4. As we have previously seen, every day has a unique energy, marked by a special frame of consciousness, which means that a day of birth is also a unique combination of consciousness polarities at different levels. It thus seems natural that the energies of our particular days of birth will shape our attitude toward the world and how we relate to our environment—that is to say, some of our personality traits. Animals are often used to symbolize the personality traits resulting from these archetypal influences. Day signs that are not directly named for an animal are linked to some totem animal, or *nagual*, as listed in figure B.2.

In addition to the cosmic tzolkin count to which all of creation is subordinated, we also have individual tzolkin counts beginning at our particular tzolkin day of birth. Thus, for instance, if you were born on the tzolkin day 8 Cauac (which is kin 99 in the cosmic tzolkin count), this would be *kin 1* in your individual count. Then the day 3 Eb *(kin 172)* is *kin* 172–99 = 73, 8 Ben, in your own individual tzolkin count. Thus, the cosmic and individual tzolkin waves run parallel, although with a constant phase difference between them. A cosmic tzolkin energy takes on an individual meaning for you depending on your own tzolkin birthday.

There are certain days in the tzolkin count when especially meaningful events may take place in your life that can serve as guidance for your life's purpose. All days ruled by your own tzolkin number and glyph, respectively, may be such days. The reason is that if you chose to incarnate in the Holy Universe of Time on a day that was 7 Ik, for instance, then all days that have the energy of either 7 or Ik will tend to reactivate your memory of why you chose to incarnate in the world and what your life's purpose is. Generally, this should mean that on days ruled by your own number and glyph your life's purpose would have some wind at its back. But, of course, such events are not necessarily always pleasant and may sometimes take the form of lessons to learn. Something similar may be true for the various tuns of the Galactic Underworld, in which someone who was born on the day 7 Ik may find that the seventh tun in this Underworld is an important time for manifestation of his or her life's purpose.

There is something else to see regarding the cosmic flow of time. Who you are and what you may become does not depend only on the energy of the day you were born. A deeper way of understanding the significance of a tzolkin day of birth is that this is the day when you enter the continuous stream of Holy Time. If, for instance, you were born on a day with the energy of 9 Ik, this by necessity means that the

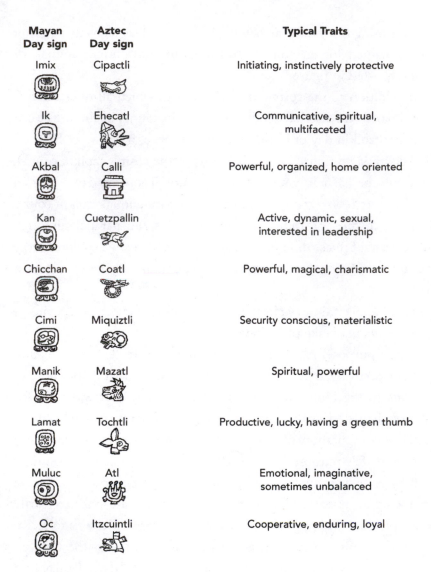

Mayan Day sign	Aztec Day sign	Typical Traits
Imix	Cipactli	Initiating, instinctively protective
Ik	Ehecatl	Communicative, spiritual, multifaceted
Akbal	Calli	Powerful, organized, home oriented
Kan	Cuetzpallin	Active, dynamic, sexual, interested in leadership
Chicchan	Coatl	Powerful, magical, charismatic
Cimi	Miquiztli	Security conscious, materialistic
Manik	Mazatl	Spiritual, powerful
Lamat	Tochtli	Productive, lucky, having a green thumb
Muluc	Atl	Emotional, imaginative, sometimes unbalanced
Oc	Itzcuintli	Cooperative, enduring, loyal

Figure B.4. Personality traits associated with the signs of birth

second day of your life was 10 Akbal in the cosmic time flow, the third 11 Kan, and so on. Through your tzolkin energy at birth you are initiated in the cosmic flow of time—a flow that will continue without interruption throughout your life. To use myself as an example, I was born on the day 5 Ix, 5 Jaguar. This means there were seven days until I experienced my first uinal shift and nine days until I experienced my first

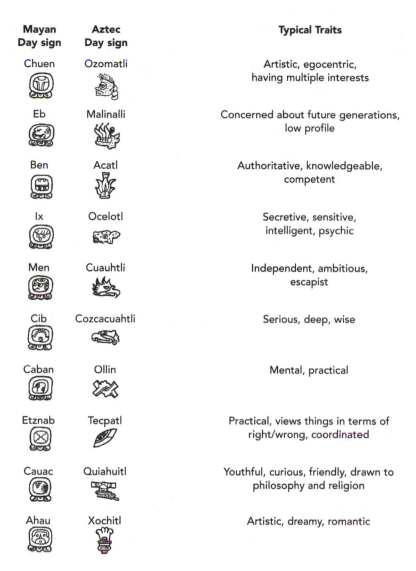

Mayan Day sign	Aztec Day sign	Typical Traits
Chuen	Ozomatli	Artistic, egocentric, having multiple interests
Eb	Malinalli	Concerned about future generations, low profile
Ben	Acatl	Authoritative, knowledgeable, competent
Ix	Ocelotl	Secretive, sensitive, intelligent, psychic
Men	Cuauhtli	Independent, ambitious, escapist
Cib	Cozcacuauhtli	Serious, deep, wise
Caban	Ollin	Mental, practical
Etznab	Tecpatl	Practical, views things in terms of right/wrong, coordinated
Cauac	Quiahuitl	Youthful, curious, friendly, drawn to philosophy and religion
Ahau	Xochitl	Artistic, dreamy, romantic

Figure B.4 *(continued)*

trecena shift, two significant early energy shifts in my life. So here I am now, many tzolkin rounds later, writing a book about the roles of the numbers 7 and 9 in creation.

Of course, it is because a cosmic plan whose light generates directed flows of activity really does exist that human beings have destinies. Some things may be meant to happen and others not, depending on whether they are or are not in line with the streams of creative light patterned by the tzolkin. Our individual flows are created by our individual tzolkin counts beginning at our day of birth. How our individual cycles interface with the cosmic cycles has a great deal to do with how our individual destinies play out and what streams of creative light will be available to us.

Not only human beings, but also everything else that is born, such as a nation (e.g., the United States, July 4, 1776 = 9 Ben), company, animal (I know a dog who, like myself, was born on the day 5 Jaguar, and we get along very well, with a deep mutual understanding), plant, or other phenomenon will generate an individual tzolkin count. Despite this potential diversity, there is still only one cosmic tzolkin count that rules all of creation, the true count, which is a harmonious overtone of all the larger cycles that rule cosmic evolution.

Another point to notice is that a certain energy in a group of people, such as a family, company, or gathering, is created depending on the tzolkin birthday energies of the participants. You can use a tzolkin chart, such as that in figure 1.10 (page 15) or figure 6.4 (page 128) and mark out the tzolkin energies of birth of members of the particular group and study the resulting pattern. Obviously, there is also much to learn about relationships between persons born with different tzolkin energies.

THE CONTEXT OF YOUR TZOLKIN DAY OF BIRTH

In the larger context an isolated tzolkin combination conveys only limited information regarding the energy of a particular day. From our studies of history it should be evident that everything is not determined solely by the particular tzolkin combination ruling our days of birth. To take a somewhat extreme example of this, someone who was born on the day 4 Eb on June 1 in the year 1 C.E. was born into an entirely different frame of consciousness, ruled by a different Heaven and Underworld, than someone born on the day 4 Eb, April 20, 2000 C.E. Thus although 4 Eb in the tzolkin count ruled both these dates, their contexts are completely different.

The day-to-day tzolkin is only a ripple on the waves of the ocean of creation; the larger waves are much more important for determining your frame of consciousness than the day-to-day tzolkin. The baktun, katun, and tun in which you were born have great divinatory significance, something that becomes evident if you study ancient Mayan stelae. What is more, if we merely consider our tzolkin days of birth, we will remain stuck in a worldview of endlessly repeating cycles, in this case 260-day tzolkin rounds. Creation, however, is both noncyclical and nonlinear. A calendar adequately describing creation needs to be both evolutionary and spiraling, and it must include several levels of evolution. It is only in such an evolutionary context, generated by taking longer cycles into account, that the energy of a tzolkin birthday gains its true meaning.

Moreover, the context of the entire tzolkin round also has great relevance to the true significance of the number of the day you were born. As mentioned, the number of your tzolkin combination means that a specific deity dominates it. The number 2 in 2 Chuen, for instance, is ruled by Tlaltecuhtli (see fig. 2.2, page 19), but the number 2 in this combination also defines that you were born in the trecena of Oc, in its second day. What this means is that you are a Chuen (Monkey) within an Oc (Dog) and that your personality has a Dog character although your day sign is Monkey. Because you were born in the second day (2) of the Dog trecena, however, your Dog character is not as strong as it would be with a number that is 12 or 13 in the same trecena. The number before a day sign, therefore, unambiguously defines in what trecena and uinal you were born, cycles that also exert influences on your personality and destiny.

Your tzolkin energy of birth is thus not only a number plus a glyph. It can be understood only from its place in the context of the whole 260-day tzolkin structure. To be born on the day sign Chuen, for instance, means something entirely different if this is 9 Chuen in the uinal of Tezcatlipoca or 2 Chuen in the uinal of Quetzalcoatl. Chuen is always at the midpoint of a uinal, but since the uinals themselves have very different energies, so will the energies of 2 Chuen and

9 Chuen. The tzolkin energy is not the number 9 plus the day sign Chuen, but the unique combination 9 Chuen. Experienced Mayan day-keepers assert that, to understand the significance of your tzolkin birthday, you have to look at it in the context of all the 260 combinations, of the total filtration pattern of light, that the tzolkin really is. It also matters in what four-day, five-day (*quintana*), thirteen-day (trecena), twenty-day (uinal), fifty-two-day, and sixty-five-day cycles you were born.

THE TUNIVERSARY

In addition to your tzolkin day of birth, you also have a birthday every 360 days that we may call the tuniversary. The tuniversary is a celebration of your birthday in the spiritual 360-day "year" rather than in the physical 365-day year. In ancient times Mayan kings would often erect stelae to celebrate the tuniversaries or katuniversaries of the days they were born. On the ancient stelae among the classical Maya, it is rare to find shaman-kings celebrating their tzolkin days of birth. What seems to have been much more important to them were their tuniversaries. The katuniversaries were of most interest because they knew the katun to be of great prophetic significance. Your tuniversary has much more to say about the ups and downs of your life, and its new beginnings, than the tzolkin day. I believe the importance of the tuniversary has been vastly underestimated in the revival of the Mayan calendar that has taken place in recent years.

To celebrate your tuniversary is a way of phasing in with the wave process of divine creation, which primarily develops according to tun-based cycles. There is a clear difference between the tzolkin day of birth and the tuniversary. While the tzolkin day reflects the energy of your birthday, the tuniversary is a nodal day in your energy cycle that has a great deal to do with to what extent your individual spiritual process is synchronized with those of certain others. Every 260 days the tzolkin energy of your birthday recurs, but every 360 days after your day of birth there is a node point in your personal tun-cycle. The two days are always marked by the same day sign, but the particular number in the thirteen-day cycle of the day you were born varies among different tuniversaries.

You may explore how your own evolutionary spiritual cycle relates to those of other people in your life by drawing a circle with 360 degrees and placing the *kin* number of your own tuniversary in that circle together with those of certain others of interest to you (fig. B.5). The resulting angles will then have something to say about how well you are synchronized with their spiritual wave movement.

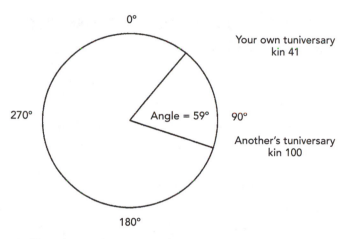

Figure B.5. A 360° circle may be used to determine angles between tuniversaries of different persons, and thus to what extent their paths of evolution are synchronized.

Thus with the tzolkin day of birth you may explore your energy compatibility with others, while through the tuniversary you may explore how much you are in sync with them, which is an entirely different matter. Angles close to 0° or 180° may be favorable to a synchronized spiritual evolution, whereas 90° or 270° might mean the very opposite.

Every thirteenth tun the two cycles, your personal tun cycle and the tzolkin, synchronize, which makes every thirteenth tuniversary an especially powerful day. At 4 × 13 = 52 tun after your birth, this power is even further accentuated. Not only are your personal 260- and 360-day cycles synchronized, they also harmonize close to, or exactly, at the day you were conceived. So the fifty-second tuniversary will be a rebirth and maybe a reconception, and hence the beginning of a new spiritual phase in your life, that of an elder.

It may also be interesting to know that for one person out of two the tzolkin day of birth coincides with his or her Gregorian birthday every forty-two solar years, meaning that the seasonal and spiritual energies of their days of birth are then brought together.

HOW TO ENHANCE YOUR SENSITIVITY TO THE CALENDAR

There are several ways of sensitizing yourself to the energies of the Sacred Calendar, to make the energies of the various days more real to you, and to learn to *tune* in to them. If you acquire a classical Mayan calendar, you should start by following the days and notice how you experience the tzolkin rounds, tun-shifts, and so on. You can learn the cosmic energies associated with the various day signs and numbers by meditating on the glyph and number of each day. You can also sensitize yourself to the energies of the calendar by creating habits or rituals (such as praying to the Day Lords) to increase your awareness of them.

You can meditate every day of the trecena to a tape playing the tone in the regular C-major scale corresponding to the particular day. Some recommend different sex on the different day signs. In the same way as the practice of t'ai chi is linked to the number 108 (explained in chapter 9), it is possible to base the practice of yoga or other spiritual exercises on the Mayan calendar. There are essences with aromas linked to all the day signs to enhance your experience of the various signs. In ancient Mexico people would also follow a cyclic twenty-day diet (fig. B.6, pages 250–251), and this could also be a way of getting into the flow. Find out what senses are most important to you—smell, sight, taste, hearing, touch, or whatever else may appeal to you—and create patterns relating either to your individual tzolkin round or to the cosmic cycle.

In the Mayan tradition specific days in the tzolkin have been assigned as good or bad for doing certain things. A few examples, taken from Marco and Marcus de Paz's *The Mayan Calendar,* are given in figure B.6. I must admit I have never regimented my own life to these kinds of recommendations, but there may in principle be a rational basis for them. Consider, for instance, how human consciousness has alternated between unitary and polarized frames of consciousness throughout the history of the Great Cycle. As a consequence, the brain's mode of operation alternates as different tzolkin combinations rule. Some days may be more suitable for creativity, while others seem to impose rest and corresponding types of activities. Nonetheless, I suspect that the recommendations in figure B.6 are overly detailed, and for most of us they probably have to be adapted to the modern world to be meaningful.

I am convinced that certain types of events do tend to happen at certain days in the tzolkin round, especially as related to the uinals, their beginning signs, their middle signs, and their ending signs. The way I use a tzolkin round is to initiate projects of my own choosing at its first uinal and then follow how the aims of these projects materialize in relation to the cosmic flow of energies that a tzolkin round entails. Such a round then always becomes a cycle of rebirth, of re-gestation, as indeed most contemporary Mayan shamans would describe it. As part of such a 260-day rebirth I am aware of the flow of uinals and how these tend to create patterns in my life in relation to the various projects I am pursuing. In this way I recognize the reality of the energy flow that the round describes, and yet exercise a free choice in the projects that I focus on.

There are many options. No one has the correct answer as to how the tzolkin is to be followed—if indeed there is one—but we may all explore it. Its energies are surely real.

Mayan Day sign	Aztec Day sign	A Good Day for . . .	Recommended Food
Imix 	Cipactli 	Buying a house	Fish
Ik 	Ehecatl 	Protecting domestic animals	Vegetable tamales
Akbal 	Calli 	Courtship	Turkey feast
Kan 	Cuetzpallin 	Praying to the spirits of mountains	Fruit of cacti
Chicchan 	Coatl 	Praying for marriage	Eggs
Cimi 	Miqiztli 	Praying for the sick	Dish made from intestines
Manik 	Mazatl 	Hunting	Deer
Lamat 	Tochtli 	The cult of corn	Rabbit
Muluc 	Atl 	Purging faults	Fish and fasting
Oc 	Itzcuintli 	Sex	Dog

Figure B.6. In the Mayan tradition, specific days in the tzolkin are considered either good or bad for certain actions or activities. A few examples from *The Mayan Calendar* by Marco and Marcus de Paz are shown here.

Mayan Day sign	Aztec Day sign	A Good Day for . . .	Recommended Food
Chuen	Ozomatli	Money, animals, and crops	Fruit
Eb	Malinalli	Praying for well-being	Fruit or herb soup
Ben	Acatl	Children	Fish
Ix	Ocelotl	Praying for rain	Turkey feast
Men	Cuauhtli	Economic well-being	Meat and beans
Cib	Cozcacuahtli	Making offerings to souls	Stewed meat
Caban	Ollin	Praying for a relative	Vegetables
Etznab	Tecpatl	Freeing people from curses	Juice fasting
Cauac	Quiahuitl	Business trips	Fruit
Ahau	Xochitl	Removing evil evocations	Turkey feast

Figure B.6 (continued)

How to Calculate Your Tuniversary and Tzolkin Day

Tuniversary Example: Suppose you were born on the Gregorian date October 3, 1960. Go to figure C.1, table 1 to find the day number corresponding to the year 1960, which is 21,915. To this you should add the number for the month of October in the tun column in table 2, which is 470, to which 3 should be added for the number of days in the month of October, giving 21,915 + 470 + 3 = 22,388. Dividing this by 360 gives 22,388/360 = 62.1888. Subtracting the integer part and multiplying the remaining decimal part by 360, you obtain 0.1888 × 360 = 68. This means that your tuniversary will always fall on the sixty-eighth day of the tun. To know the date your spiritual birthday falls on in a given year, begin at the first day of a new tun (bottom row of fig. 7.1, page 138—for instance, December 10, 2003), and count until you come to day number 68, which will take you to February 15, 2004. Of course, from year to year the Gregorian date corresponding to your tuniversary will vary.

Year	Day	Year	Day	Year	Day	Year	Day
1900	0	1928	10227	1956	20454	1984	30681
1901	365	1929	10592	1957	20819	1985	31046
1902	730	1930	10957	1958	21184	1986	31411
1903	1095	1931	11322	1959	21549	1987	31776
1904	1461	1932	11688	1960	21915	1988	32142
1905	1826	1933	12053	1961	22280	1989	32507
1906	2191	1934	12418	1962	22645	1990	32872
1907	2556	1935	12783	1963	23010	1991	33237
1908	2922	1936	13149	1964	23376	1992	33603
1909	3287	1937	13514	1965	23741	1993	33968
1910	3652	1938	13879	1966	24106	1994	34333
1911	4017	1939	14244	1967	24471	1995	34698
1912	4383	1940	14610	1968	24837	1996	35064
1913	4748	1941	14975	1969	25202	1997	35429
1914	5113	1942	15340	1970	25567	1998	35794
1915	5478	1943	15705	1971	25932	1999	36159
1916	5844	1944	16071	1972	26298	2000	36525
1917	6209	1945	16436	1973	26663	2001	36890
1918	6574	1946	16801	1974	27028	2002	37255
1919	6939	1947	17166	1975	27393	2003	37620
1920	7305	1948	17532	1976	27759	2004	37986
1921	7670	1949	17897	1977	28124	2005	38351
1922	8035	1950	18262	1978	28489	2006	38716
1923	8400	1951	18627	1979	28854	2007	39081
1924	8766	1952	18993	1980	29220	2008	39447
1925	9131	1953	19358	1981	29585	2009	39812
1926	9496	1954	19723	1982	29950	2010	40177
1927	9861	1955	20088	1983	30315	2011	40542

Table 1

Month	Tun	Tzolkin
January	197	237
January (if leap year)	196	236
February	228	268
February (if leap year)	227	267
March	256	296
April	287	327
May	317	357
June	348	388
July	378	418
August	409	449
September	440	480
October	470	510
November	501	541
December	531	571

Figure C.1. Table 1 shows day numbers for different Gregorian years of birth (from Jenkins, Tzolkin). **Note:** Gray shading is included only for readability and is not related to Days and Nights. Table 2 shows day numbers corresponding to different months; to be used to calculate the tun (or tzolkin) birthday.

Table 2

Calculating a Spiritual Birthday (Tuniversary)

1. The day number of the year (from fig. C.1, table 1) is:
2. The day number of the month (from fig. C.1, table 2) is:
3. The specific day in the month is:
4. The sum of (1), (2), and (3) is:
5. This sum divided by 360 is:
6. The decimal part of this number multiplied by 360 is:
7. To determine your tuniversary for a particular year, count this number of days from any of the tun beginning dates in the bottom row of figure 7.1 on page 138.

Tzolkin day example: Suppose you were born on January 19, 1960, and want to know what day this corresponds to in the tzolkin. Add 21,915 + 236 (1960 was a leap year) + 19 = 22,170, from figure C.1. Divide this number by 260, giving 22,170/260 = 85.2692. Subtract the integer part and multiply by 260, which gives 0.2692 × 260 = 70. Your *kin* number is thus 70 in the tzolkin. In figures 1.9–1.10 (pages 14–15), you find that this corresponds to 5 Oc.

What time to consider as the beginning of the day is an issue that may not yet have been adequately solved. It is my own experience, however, that those born before sunrise should be placed under the day sign of the previous day.

As an alternative to this tedious procedure, a calculator for determining tzolkin days and tuniversaries may be downloaded from my Web site, www.calleman.com.

The War of the West Against Iraq

On the day 1 Men (1 Eagle) in the Mayan tzolkin count, in the fifth uinal of the fifth tun (ruled by the serpent of duality) of the Galactic Underworld, the Western forces of the United Kingdom and the United States attacked Iraq and defeated the Saddam Hussein regime about two trecenas later. The invasion has extensive ramifications, and it is already evident that its consequences on global politics have been very different from—and largely opposite to—those of the first Gulf War. The reason is that this attack took place in an entirely different context of the universe of time described by the Mayan calendar. This war occurred in the Third Day of the Galactic Underworld.

In 1991, in contrast, the Seventh Day of the Planetary Underworld was rapidly approaching (beginning on February 11, 1992). Hence the 1991 war came to be a rallying point for a new planetary consensus in which democratic systems of rule seemed to be emerging everywhere, in parallel with the former division of the world into an Eastern and a Western block coming to an end. As this Underworld's Seventh Day began in 1992, the Internet came into existence, and it seemed that all boundaries preventing networking and communication on a global scale would disappear. A year earlier an essentially unanimous global community led by the United States launched a successful military offensive against Iraq, in response to its attempt to forcibly reincorporate Kuwait. This campaign seemed to unify the whole world. As a result, the United States emerged at the beginning of the Seventh Day of this Underworld as the world's only superpower. In the ensuing period of unprecedented economic growth, the United States seemed to become all-powerful.

The 2003 war against Iraq, however, occurred in a global setting whose energies were very different from those during the earlier war. We are now increasingly dominated by the dualist consciousness that has been carried by the Galactic Underworld that began on January 5, 1999. The recent largely unprovoked offensive by the Western coalition thus occurred in a different context of consciousness—a context that may easily divide nations belonging to the East and the West rather than unify them. The light now falls on the Eastern Hemisphere and the right-brain hemisphere, and in the preparatory phase leading up to this war, especially from the onset of the Third Day of the Galactic Underworld on December 15, 2002, it became evident that the times are not the same as in 1991. Western wars are no longer unanimously applauded. This is fundamentally because the Galactic Underworld brings a different frame of consciousness, with new values, to the world. Consequently, the key to understanding the implications of this war lies in understanding this new frame of consciousness.

Many tend to view the 2003 attack as the product of a particular American administration with a particular president, or as a reaction to the terrorist attacks of September 11, 2001. But the reasons for the second Gulf War actually go much deeper. To understand these reasons, it might be helpful to read my previous book, *The Mayan Calendar* (Garev, 2001), which described the divine time plan and the evolution of consciousness as understood from the Mayan calendar. Ultimately, the only way to evaluate a prophetic tool is to see how accurate it has been in the past for predicting the future. While I must admit that I made some mistakes in details, it is indisputable that the basic subdivisions of the world that I proposed, in terms both of space and time, have been verified to the fullest.

I will repeat some of the general predictions I had already made regarding the Galactic Underworld and how it would develop. One such basic notion that has now been verified is the division of the world into three major sections: the West, the East, and continental Europe in between. The final war council on the Portuguese islands of the Azores was held between Spain, the United States, and the United Kingdom. These are the nations that are located entirely west of the galactic resonance unit, as this was delineated in figure 78 in my earlier book. It is obvious that the recent war against Iraq was a war of the West.

This overall view of a tripartite world is further substantiated by the fact that nations such as Germany, France, Belgium, and Sweden spoke out against this war. The consciousness of these nations is strongly influenced by the presence of the planetary midline, and they are thus not Western in the narrow sense of the word; in the future they will appear even less so. In the Galactic Underworld the West has, in other words, become more directly related to the World Tree and has become more or less synonymous with Anglo-Saxon nations. Germany was one of the nations whose population most vehemently opposed the war—a fact that is related to Germany's location directly under the World Tree. While some thirty nations are said to have supported the recent war, most Eastern nations, including Russia and China, were opposed to it. What we see now is the beginning of an irreversible process of divorce between Europe and the West, which would have been unthinkable in the Planetary Underworld.

The recent war also occurred at a definite point in the cosmic time plan, the Third Day of the Galactic Underworld. I had predicted that this whole Underworld, after the relative absence of conflicts in the period 1992–1999, would lead to an intensification of conflicts between East and West. This may seem obvious today, as these conflicts have escalated immensely, but it was certainly not clear in the 1990s, when the Cold War had come to an end, together with much of the tension in the Middle East. The prediction that it would be the Days of the Galactic Underworld that would most likely give rise to such conflicts and wars between East and West has also been verified. As such periods begin, dominated by a new yin/yang duality of consciousness, a shift in the balance of forces easily arises, generating tension that may result in warfare.

In the United States, which is still favored by the yin/yang duality of the National Underworld, it is traditional that a majority of people will support their president when he decides the country must go to war. Thus many who do not directly benefit economically from a war—for instance, from the defeat of Iraq—still support military campaigns. Many of those who are influenced by dualist thinking will seek to remain in the light created by this duality (see fig. 3.11a, page 46) and will back the self-proclaimed "right" of its president and administration (defined as "good") to attack countries that are outside its control (defined as "evil"). Because the West has been favored by the yin/yang duality in the consciousness ruling until now, it is also the West that will be the last to surrender such dualist thinking. This is the mind-set that for a long time created the basis for the military power of the West.

Now, if the emerging Galactic Underworld favors the Eastern Hemisphere, why did the West once again emerge victorious against a nation of the East? First, there was never much reason to expect that Iraq would be able to challenge the West on its own terms, that is, in technological or military strength. Second, it is also obvious that the brutal Saddam regime lacked support among its own people.

Thus it is not only Western world dominance that will come to an end with the Galactic Underworld; dominance and inequality as such will come to an end. Glaring local dictatorships, such as that of Saddam Hussein, will also come to an end in a process parallel to the global dominance of the West, although the detailed course of events is unforeseeable (at least to me). Ultimately, this is a process brought about by a change in consciousness. Dominance is a mental attitude that was generated by the duality of National Underworld, and so, as the Galactic Underworld progresses, it will disappear everywhere.

Hence the withering of hierarchies and dominance is partly what the Galactic Underworld is about, and with every Day this process will be reinforced. In Night Five, ruled by Tezcatlipoca, the lord of darkness (essentially the Gregorian year 2008), we will see the last desperate, and at the same time most forceful, attempt to secure control by the forces seeking to maintain dominance.

To see how the current Day has strengthened the East and the holistic right-brain hemisphere, consider the shift among Europeans, who traditionally have been considered pro-Western. The period prior to the second Gulf War saw manifestations for peace of an unprecedented scope all over the world, but especially strongly in central Europe, where people from all walks of life, regardless of political views, came out for peace. Although many Eastern nations, such as Russia and China, also opposed the war, the shift was probably the greatest in central Europe. This desire for peace represents a great change and is an aspect of what the Galactic Underworld is carrying and will continue to carry. In the emerging consciousness of the Galactic Underworld there is either peace in the present or not at all, and in its coming Days this insight will become increasingly widespread.

While the West has been victorious in military terms, its credibility in much of the world has decreased enormously—partly because it defied international law to launch its offensive, by ignoring the decisions of the Security Council of the United Nations; and partly because the justifications advanced for the war turned out to be nonexistent or even concocted. The erosive damage done to the United Nations is such that this organization will probably never recover as a meaningful political forum or potent world body politic. In terms of legitimacy, the West seems to have sawed off the branches it has been sitting on, and it seems obvious that in the event of another military campaign the opposition among people in other countries will be even greater. But again, the reason that world opinion has changed so much in its relation to the United States is that the spiritual winds of the cosmic plan are now blowing in a new direction. These winds are beyond human manipulation and invulnerable to weaponry.

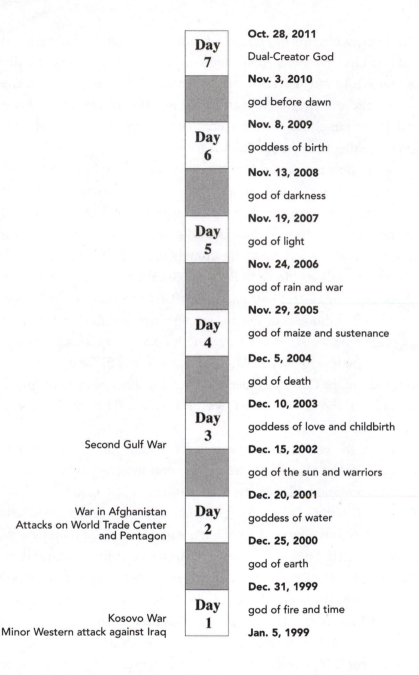

Day 7		Oct. 28, 2011 Dual-Creator God
		Nov. 3, 2010 god before dawn
Day 6		Nov. 8, 2009 goddess of birth
		Nov. 13, 2008 god of darkness
Day 5		Nov. 19, 2007 god of light
		Nov. 24, 2006 god of rain and war
Day 4		Nov. 29, 2005 god of maize and sustenance
		Dec. 5, 2004 god of death
Day 3		Dec. 10, 2003 goddess of love and childbirth
Second Gulf War		Dec. 15, 2002 god of the sun and warriors
War in Afghanistan Attacks on World Trade Center and Pentagon	**Day 2**	Dec. 20, 2001 goddess of water
		Dec. 25, 2000 god of earth
		Dec. 31, 1999 god of fire and time
Kosovo War Minor Western attack against Iraq	**Day 1**	Jan. 5, 1999

Figure D.1. The Galactic Underworld and its Days and Nights with the ruling deities and a timeline of Western warfare against the East. The first Gulf War started at the end of the Sixth Night of the Planetary Underworld, when each of the thirteen Heavens were 19.7 years. In the Galactic Underworld each Heaven equals 360 days. Note that the actions took place during the Days.

The struggle between these two conflicting forces, on one hand the force striving to maintain the world dominance of the West, and on the other that striving to unify the East and the West, has already had very significant results in terms of world politics. The wheeling and dealing preceding the war—with the West trying to buy votes from poorer countries by offering loans and financial support—and the threat after the war of consequences for France for having opposed it has even further exposed the kind of "democracy" that is operating on the global level. The paradox is that the leader of the United States, elected by a minority of about 50 million of his people, reserves the "right" to go to war anywhere in a world of more than 6 billion that certainly did not elect him. Globally speaking, this looks like the kind of "democracy" that existed in many places in the nineteenth century, when one had to earn a certain income to be eligible to vote.

How long will the current war in Iraq last? This all depends on what we mean by *the current war*. It is probably accurate to say that since the beginning of the Galactic Underworld in early 1999 we have been living in World War III, and this will probably go on in varying forms at least until early 2009. However, World War III does not have the same character as World Wars I and II that occurred in the Planetary Underworld, and thus it may not be recognized as such. Even if we look more specifically at the current occupation of Iraq, it seems doubtful that it will come to an end. The dilemma of the Western coalition is obvious: any puppet regime that it may install to safeguard its interests would immediately fall when the coalition pulled out its troops. The United States cannot easily withdraw from its occupation of Iraq, so it is likely that in the Fourth and Fifth Days of the Galactic Underworld, Iraq, as well as Israeli-Palestinian tensions, will remain a focus of East-West conflicts, which very likely will expand to encompass the whole region.

The symbolic importance of the locations and nations involved in this drama should not escape us. If anyone doubted, the Galactic Underworld is the Apocalypse, in which the Beast of dominance will be cast down from the heavens. The Book of Revelation refers many times to Jerusalem and Babylon, and this war surely is part of the scenario of the end times. It is no accident that, as the Long Count is coming to an end, it is acted out in the very places where its first expressions arose. The modern system of global dominance is confronting what became of the earliest forms of patriarchal fundamentalist dominance. The Eagle of dominance now returns to crush the egg it came from, but if it does, will it survive? The serpent of duality tries to bite its own tail thinking it is prey, but if it does, how and where will it be able to move? The current conflicts in the Middle East bring the alpha to the omega, where the modern world sees itself reflected in the mirror of its past. Human civilization has come to the end of the road. It is time to recognize that we are all One.

Glossary

18-Rabbit: Mayan shaman king ruling in Copan

acupuncture: traditional Chinese branch of medicine in which needles are inserted at certain meridian points

Ahau: twentieth day sign of the uinal, meaning "light/lord"

ahauob: Mayan rulers (lords)

Ah-Cacaw: ruler in Tikal in the seventh century C.E.

alautun: period of 63.1 million years

alpha waves: brain waves in the frequency range 8–13 Hz

amino acids: components of proteins

An: Sumerian god of heaven

anthropology: study of the development of human beings

apocalypse: Greek word meaning "revelation," usually referring to the scenario described in the Book of Revelation

Arbenz, Jacobo: Guatemalan president working for reforms who was ousted by the CIA in 1953

Artaxerxes III: Persian king ruling in the late fourth century B.C.E.

ascension: theological term meaning "rising"

Ask: first human being in Norse mythology, also meaning "ash tree"

astroarchitects: architects basing their constructions on alignment with the celestial bodies

astrology, Babylonian: zodiacal astrological system based on the positions of the celestial bodies developed in ancient Babylon; what in today's world is commonly identified as astrology

astrology, Mayan: system of divination developed by the ancient Maya; based on the inherent spiritual energies of time

Atl: Aztec name of the ninth day sign of the uinal, meaning "water"

Atlantis: ancient mythological continent

avatar: Indian term for the incarnation of an enlightened soul

Aztec: name commonly used for the Mexicas, based on their origin in Aztlan

Bacabs: directional gods of the Mayans

baktun: period of 394 years

Beast: main evil force described in the Book of Revelation

Ben: thirteenth day sign of the uinal, meaning "reed"

beta waves: brain waves in the frequency range 13–40 Hz

Big Bang: beginning of the creation of the universe; a giant "explosion" in which matter was first formed from light some 15 billion years ago

biosphere: region of the earth's crust and atmosphere harboring living organisms

Brahma: God of Creation in the Vedic religion

Bretton Woods system: financial system of the post–World War II era outlined at a meeting in Bretton Woods, New Hampshire, in the United States in 1944

Burner Days: set of days that are specially celebrated; they occur every 65 days of the tzolkin

Calendar Stone, Aztec: calendar stone of Aztec origin kept in the Museum of Anthropology in Mexico City; one of the most common symbols of the nation of Mexico

capitalism: economic system whose driving force is the accumulation of profits

celestial: pertaining to the heavens

Cellular Creation Cycle/Cellular Underworld: first level of creation of thirteen hablatuns, beginning 16.4 billion years ago

center of the world: Yaxkin, the Mayan center of the Four Directions

Chac: Mayan god of rain

chakra, crown: chakra centered on the top of the head

chakra meditations: meditations to purify the chakra system

chakra system: Hindu system dealing with the organizing structure of the energy flow of the human being

Chalchiuhtlicue: Aztec goddess of water, ruler of the third day of the trecena

Chaldea: biblical name of Sumer

channeling: receiving information from a higher spiritual source

Cherokee: Native American people originally living in present-day Georgia, later displaced to reservations in Oklahoma

Chicchan: fifth day sign of the uinal, meaning "serpent"

Chichén Itzá: major Mayan temple site on the northern Yucatán peninsula

chikin: Mayan name for the western direction

Chuen: eleventh day sign of the uinal, meaning "monkey"

Cib: sixteenth day sign of the uinal, meaning "vulture/owl"

Cimi: sixth day sign of the uinal, meaning "death"

Cinteotl: central Mexican god of maize (corn) and sustenance, ruler of the seventh day of the trecena

Cipactli: Aztec name for the first day sign of the uinal, meaning "alligator"

classical count: tzolkin count used in classical times by the Maya; still in use among the Quiché-Maya

classical Maya: Maya who lived in classical times, about 200–850 C.E.

Coatl: Aztec name for the fifth day sign, meaning "serpent"

codex: ancient Mesoamerican book, usually made from bark

Codex Borgia: late postclassical Aztec codex

Codex, Dresden: postclassical Mayan codex used by the German librarian Förstemann to elucidate the calendrical system of the Maya

Confucius: ancient Chinese philosopher (c. 550 B.C.E.)

conquistadors: Spanish soldiers who conquered the Americas

consciousness: invisible boundary that separates inner and outer reality

cosmic consciousness: all-encompassing consciousness

Cortés, Hernán: Spanish conqueror of Mexico

cosmic mirror: opportunity for humankind to be reflected by the cosmos

cosmic projector: universal framework for the projection of archetypal information dominating the various Mayan time cycles

cosmic pyramid: nine-story pyramid

cosmopolitan: not belonging to a specific nation

count of days: another term for the tzolkin

creation field: energy field organizing the divine process of creation

Cross, Heavenly: World Tree; it exists as holographic projections on the level of the galaxy, the earth, and the individual

Cross, Invisible: another term for Heavenly Cross

Cross of Vuotan: ancient Norse name of the Sun Wheel, identical with the Medicine Wheel

crust: surface layer of the earth

Cuauhtli: Aztec name for the fifteenth day sign, meaning "eagle"

Cuetzpallin: Aztec name for the fourth day sign, meaning "lizard"

cuneiform: type of writing used in ancient Mesopotamia

Curry and Hartmann lines: lines in the earth's energy grid sensed by dowsers

cyber money: funds that exist only in electronic form, such as in bank computers

Dark Ages: name given to the early medieval period, the centuries following the fall of the Roman Empire

day keepers: special Mayan persons responsible for keeping track of the passage of days

Day Lord: another name for day sign

day signs: the names of the deities ruling the days of the uinal

December 21, 2012 C.E.: end date of the Long Count

delta waves: brain waves in the frequency range 1–4 Hz

Deutero-Isaiah: author of the later sections of the Book of Isaiah in the Bible, promoting a view of God as universal

divine light: energy of awareness serving to project archetypal information on creation

divine plan: the exact progression of spiritual energies determining the evolution of consciousness

dowser: person with the ability to sense the earth's radiation

Dreamspell: calendar inspired by the Maya, invented by José Argüelles in 1990

ecclesiastical: phenomenon pertaining to the church

EEG: electroencephalogram; an instrument that measures brain waves

Ehecatl: Aztec name for the second day sign of the uinal, meaning "wind"

Eightfold Path: Buddhist teaching as to how to attain the enlightened state

enlightenment: liberation from domination of lower aspects of self; realization of unity with higher Self

Enlightenment, the: movement promoting rational thought and anticlericalism in Europe in the mid-eighteenth century

epiphysis: pineal gland, monitoring light and linked to the "third eye"

equinox: one of two days in the year when the day and the night are equally long

esoteric: knowledge of the inner spiritual reality

ethereal: nonmaterial

Etznab: eighteenth day sign of the uinal, meaning "flint/obsidian"

evening star: Venus appearing in the evening sky

extrasolar planets: planets around other stars than our own sun

Familial Creation Cycle/Familial Underworld: third level of creation, developed by thirteen kinchiltuns beginning 41 million years ago

Feathered Serpent: main Mesoamerican deity, symbolizing light and the creation of culture and civilization; also called Quetzalcoatl or Kukulcan, the Plumed Serpent

Federal Reserve (the Fed): National Treasury of the United States, instituted in 1913

First Father: Mayan deity that raised the World Tree; also known as the maize god

Five Worlds: prophetic idea common among Native American peoples related to dividing the tzolkin into five different but equal segments

Four Worlds: prophetic idea common among Native American peoples related to dividing the tzolkin into four different but equal segments

free will: philosophical idea that human beings are in a position to make choices that are not predetermined

Gaia meditation: meditation designed to enhance resonance with the earth

Galactic Creation Cycle/Galactic Underworld: eighth level of creation, developed by thirteen tuns beginning January 5, 1999

galaxy: star system; sometimes, as in the case of the Milky Way, spiral shaped and with hundreds of billions of stars

gamma waves: brain waves in the frequency range 30–90 Hz

Genghis Khan: Mongol ruler in the early thirteenth century who founded the Mongol Empire

German-Roman Empire: empire founded in 962 C.E. by Otto the Great in Germany; it was consecrated by the pope as a continuation of the Roman Empire and was finally dissolved in 1806

global brain: organization of the planet according to the functionalities of the human brain

global chakra system: planetary meridians of ethereal energies organized in relation to the World Tree

glyph: symbol of Mayan writing

Good Friday: day celebrated as the crucifixion of Christ and the day of the landing of Cortés in Mexico

Great Cycle: creation cycle that began June 17, 3115 B.C.E., and ends October 28, 2011

Gregorian calendar: calendar of physical time instituted by Pope Gregory XIII in 1582 and currently in use on a worldwide scale

haab: Mayan 365-day cycle

hablatun: period of 1.26 billion years

Harmonic Convergence: spiritual celebration (August 16–17, 1987) on the traditional tzolkin days 1 Imix and 2 Ik

hemisphere: half-sphere

Hermetic principle: principle attributed to the so-called Hermetic tradition, an ancient esoteric school in Egypt

Hero Twins: Hunahpu and Xbalanque, chief mythological characters of the Popol Vuh, the Bible of the Maya

hierarchical: arranged or organized in a graded or ranked series of different levels

hieroglyph: basic sign of ancient Egyptian writing

Hindu: referring to the dominant religion of India, which emphasizes dharma

holocaust: destruction or annihilation; major catastrophe

holographic: relating to a structure in which the whole is microscopically reflected in each of its parts

holographic resonance: relay system for synchronistic information transmission from the macrocosmos to the microcosmos

Homo habilis: first species of the *Homo,* living about 2 million years ago in East Africa

Hosea: one of the great Jewish prophets

Huehuecoyotl: Aztec god of dance, ruler of the fourth day of the uinal

Hunahpu: one of the Hero Twins (One Ahau) of the Popol Vuh

Huns: nomadic people of Asian origin in the early medieval period

hypophysis: a central organizing gland of the endocrine system in the mammalian brain

hypothalamus: a central organizing gland of the endocrine system in the mammalian brain, closely linked to the hypophysis

Hz (Hertz): measure of frequency (per second)

Ik: second day sign of the uinal, meaning "wind"

Imix: first day sign of the uinal, meaning "alligator"

In Lak'ech: Mayan word of greeting meaning "I am another yourself"

In Lak'ech philosophy: Mayan philosophy that we are one with all living things

inner core: region of the earth's center

ionosphere: ionized region of the atmosphere reflecting radio waves

Isaiah: Jewish prophet in the mid-sixth century B.C.E.

Itzcuintli: Aztec name of the tenth day sign of the uinal, meaning "dog"

Ix: fourteenth day sign of the uinal, meaning "jaguar"

kalabtun: period of 160,000 years

Kalki: Indian avatar aiming at the enlightenment of humanity

Kan: fourth day sign of the uinal, meaning "lizard," "seed," or "net"

karma yoga: Hindu phrase meaning "path of action" or "doing good"

katun: period of 19.7 years (7,200 days)

kin: sun, period of one day

kinchiltun: period of 3.2 million years

Kukulcan: Mayan name for the Feathered (or Plumed) Serpent, called Quetzalcoatl in Nahuatl, the language of the Mexica

Lacandon: Mayan group living traditionally in the rain forest around the River Usumacinta

Lakin: Mayan deity ruling the East

Lamat: eighth day sign of the uinal, meaning "rabbit"

Lao-tzu: Chinese philosopher living around the midpoint of the sixth century B.C.E.

leap day: day introduced in February every fourth year in the Gregorian calendar

ley lines: lines identified by dowsers as part of the earth's ethereal body, usually related to human spiritual traditions

lintel: Mayan stone carving with inscription

Long Count: creation cycle that began August 11, 3114 B.C.E., and ends December 21, 2012

macrocosmos: large-scale manifestations of the cosmos

Mahavira: founder of the Jain religion

Malinalli: Aztec name for the twelfth day sign of the trecena, meaning "grass"

mammalian brain: the lateralized brain structure of mammals; its functions differ between the two hemispheres

Mammalian Creation Cycle/Mammalian Underworld: second level of creation, developed by thirteen alautuns beginning 820 million years ago

Manik: seventh day sign of the uinal, meaning "deer"

mantle: region between the earth's crust and its core

Mauna Kea: volcano on the Big Island of Hawaii; it is a major center for astronomical observations

Mayahuel: Aztec goddess of *pulque* (a Mexican alcoholic beverage); ruler of the eighth day of the uinal

Mayapan: postclassical Mayan ceremonial center in the Yucatán

Mazatl: Aztec name of the seventh day sign of the uinal, meaning "deer"

Medicine Wheel: wheel that includes the Four Directions, used for ceremonial purposes by North American Indians

Men: fifteenth day sign of the uinal, meaning "eagle"

menorah: seven-armed candelabra symbolic of creation in the Judaic tradition

Mesoamerica: "Middle America," archaeological term for the cultural area ranging from northern Mexico to Honduras

metaphor: verbal image used to illustrate an abstract phenomenon

Mexica (also known as Aztec): people who settled in central Mexico in the fourteenth century

Mixtecs: Mesoamerican people inhabiting the region of Oaxaca at the time of the Spanish Conquest

Mongol Storm: attack of Mongolian nomads across the Eurasian continent in the early thirteenth century

morning star: Venus appearing in the morning sky

Mosaic Law: law of Moses

Mount Kailas: holiest of Tibetan mountains; considered as the center of the universe

Muluc: ninth day sign of the uinal, meaning "water"

Nagual: sometimes a totem; a symbolic manifestation of the Otherworld

Napoleonic Wars: wars fought by the French during the rise and fall of Napoleon, 1794–1814

National Creation Cycle/National Underworld: sixth level of creation, developed by thirteen baktuns with a total duration of 5,125 years

New Age: modern movement prophesying the advent of a New Age

new economy: economy generated by the information technology revolution

New Jerusalem: new world emerging at the end of time as described in the Book of Revelation

nirvana: enlightened state in the Buddhist tradition

nodal point: point where a sinusoid curve crosses the axis

number 7: holy number in Mesoamerica, the number of light pulses in each Underworld

number 9: holy number in Mesoamerica, the number of Underworlds

number 13: holy number in Mesoamerica, the number of Heavens

number 108: holy number in the Hindu and Buddhist traditions; Indian sages often take this as one of their names

Oaxaca: region in southern Mexico

obsidian: volcanic glass

Oc: tenth day sign of the uinal, meaning "dog"

Oceania: collective name for lands in the Pacific Ocean

Ocelotl: Aztec name for the fourteenth day sign, meaning "jaguar"

octahedral: having the shape of an octahedron, one of the Platonic bodies with eight equivalent triangular faces

October 28, 2011: end date of the divine creation cycles

Ollin: Aztec name for the seventeenth day sign, meaning "movement"

Omecinatl: female aspect of the supreme Aztec deity of creation

Olmecs: ancient Mesoamerican people emerging around 1500 B.C.E. who inhabited the gulf coast of Mexico

Ometeotl: male aspect of the supreme Aztec deity of creation

optimal: best under a given set of circumstances

oscillation: wave movement

outer core: region between the earth's inner core and its mantle

overtone: tone of higher frequency generated by resonance with a basic tone

Oxlaj, Alejandro: Quiché-Maya elder

Ozomatli: Aztec name of the eleventh day sign, meaning "monkey"

Pacal: king of Palenque in the seventh century C.E.

papacy: institution of the pope

passage: movement (transit) by a planet across the disc of the sun

Patecatl: Aztec god of medicine ruling the twelfth day sign of the uinal

patriarchal: relating to domination by a male figurehead

patriarchs: early biblical leaders; Abraham, Isaac, and Jacob

Paul: apostle who played a key role in the early dissemination of the Christian faith

Pharaoh Djoser: one of the early pharaohs of Egypt; the first major pyramid of that culture was dedicated to him

physical biology: discipline consisting of the interaction of physics with biology

physical time: notion of time as being based on the cyclical movement of material bodies

pictun: period of 7,900 years

Planetary Creation Cycle/Planetary Underworld: seventh level of creation, developed by thirteen katuns beginning in 1755 C.E.

planetary midline: vertical arm of the World Tree, corresponding to longitude 12° East on the earth

Pleiades: group of stars in Taurus, part of Gould's Belt

Plumed Serpent (or Feathered Serpent): called Kukulcan by the Yucatec Maya and Quetzalcoatl by the Aztec

polarity: duality creating tension

polytheist: referring to religions in which many deities are worshipped

Popol Vuh: "Book of Advice," text of ancient origin relating the Mayan creation mythology, sometimes called the Bible of the Maya

postclassical Maya: Maya living in the period 850–1250 C.E.

prana: Sanskrit word for the energy of life

precession: circular movement of the earth's axis

prophecy: prediction about the future based on resonance with spiritual domains

Protestantism: Christian movement emerging in the sixteenth century based on specific protests against the Catholic Church

Pythagoras: Greek thinker living in Syracuse (Sicily) in the mid-sixth century B.C.E. who studied the harmonies of numbers

qi: life force in Chinese philosophy

quantum mechanics: physics of wave/particle duality

Quetzal: the national bird of Guatemala, bearing long, shining feathers

Quetzalcoatl: Aztec (Nahuatl) name for the Plumed (or Feathered Serpent) called Kukulcan by the Yucatec Maya

Quiahuitl: Aztec name for the nineteenth day sign of the uinal, meaning "rain"

Quiché-Maya: group of Maya living in present-day Guatemala

quintana: five-day period

Qur'an (Koran): holy scripture in the Islamic religion, consecrated by the Prophet Muhammad in 632 C.E.

Reformation: movement favoring a reform of the Catholic Church; founded by Luther, Calvin, and others

Regional Creation Cycle/Regional Underworld: fifth level of creation of thirteen pictuns, beginning 102,000 years ago

reincarnation: idea that the human soul passes from life to afterlife to life again

Renaissance: cultural movement expressing the liberation of the individual, especially artistically, in late medieval Europe

Revelation: Apocalypse, the end scenario of creation in the Bible

Romanticism: cultural movement in the early nineteenth century in Europe

rotation frequency: number of rotations per second

Sacred Calendar: name given to the 260-day tzolkin

Sacred Days of Venus: day signs on which Venus may emerge as a morning star

Saint Francis of Assisi: Italian saint in the early thirteenth century, founder of the Franciscan order

Schumann resonance: electromagnetic impulses in the cavity between the ionosphere and the earth's crust

shaman: person with the ability to contact and see into the Otherworld

Shiva: Hindu god of creation and destruction

Six Sky Lord: deity ruling the sixth Underworld

solar wind: wind of charged particles moving from the sun toward the earth

solar year: period corresponding to one revolution of the earth around the sun

spinning jenny: automatic spinning machine invented in 1794

spiritual bodies: ethereal, nonphysical aspect of the body

spiritual energy: life force; prana or qi

spring and autumn equinoxes: days of spring and autumn, respectively, when the duration of the day equals that of the night

stele (or stela): stone slab erected to celebrate a ruler (plural: stelae)

Stone Age: era in human evolution before metals came into use

Stonehenge: arrangement of ancient stone megaliths in southwestern England, near Bath

subconscious awareness: an awareness that is unknown to the conscious mind

Sumer: earliest higher civilization in ancient Mesopotamia, present-day Iraq

Sun Wheel: symbol showing a perpendicular cross in a circle

superstition: belief without foundation

sutra: verse or precept (or a collection thereof) from the Buddhist or Vedic scriptures

synchronicity: statistically unlikely event that appears meaningful and meant to happen

Taoism: Chinese philosophy developed by Lao-tzu around the middle of the sixth century B.C.E.

Tecciztecatl: Aztec goddess of the moon, ruler of the sixth day sign of the uinal

Tecpatl: Aztec name for the eighteenth day sign of the uinal, meaning "obsidian knife"

Teotihuacán: major ceremonial pyramid complex and commercial center outside of today's Mexico City; it flourished from about the time of the birth of Christ to the early eighth century C.E.

Teotihuacanos: ancient Mesoamerican people living around Teotihuacán

Tepeyollotl: Aztec deity, the heart of the mountain, ruler of the third day sign of the uinal

Tezcatlipoca: Aztec god of darkness, ruler of the tenth day of the trecena and the thirteenth and fifteenth days of the uinal; the nemesis of Quetzalcoatl

theta waves: brain waves in the frequency range 4–7 Hz

Thirty Years' War: religious war raging in Europe between Catholics and Protestants in the period 1618–1648

Tikal: major Mayan temple site in present-day Guatemala

time lock: factor blocking an aspect of consciousness for a specific period of time

Tlacolteotl: Aztec goddess of love consuming the filth of men, and ruler of the fifth day of the trecena

Tlahuizcalpantecuhtli: Aztec deity associated with the morning star and ruling before dawn, ruler of the twelfth day of the trecena

Tlaloc: Aztec god of rain and war, ruler of the eighth day of the trecena

Tlaltecutli: Aztec god of the earth, ruler of the second day of the trecena

Tochtli: Aztec name of the eighth day sign, meaning "rabbit"

Toltecs: Mesoamerican people whose capital was Tula in the state of Hidalgo

Tonacatecuhtli: Aztec god of procreation, ruler of the first day of the uinal and (sometimes) the seventh day of the trecena

Tonalaciuatl: female aspect of Tonacatecuhtli

Tonalpouhalli: Nahuatl (Aztec) name for the tzolkin chart

Tonatiuh: Aztec god of the sun and warriors, ruler of the fourth day of the trecena and the nineteenth day of the uinal

trecena: thirteen-day period of the tzolkin; a round of tones 1–13

Tribal Creation Cycle/Tribal Underworld: fourth level of creation of thirteen kalabtuns beginning 41 million years ago

True Count: tzolkin count used by the Maya for 2,500 years

True Cross: World Tree and its projections on the galactic and planetary levels

tun: period of 360 days

tuniversary: spiritual birthday occurring every 360 days

Turtle Island: Native American name for North America proper

tzolkin: 260-day count, also called the Sacred Calendar

uaxaclahunkins: Mayan name for an eighteen-day cycle

uinal: twenty-day period; a round of the twenty Day Lords

uinic: Mayan word for human being

Universal Creation Cycle/Universal Underworld: ninth level of creation of 260 days beginning February 11, 2011

universal human being: unlimited human being with a cosmic consciousness developed by the Universal Underworld

Venus cycle: an approximately 584-day progression through the phases of Venus as seen from the earth

Venus passage: passage (transit) of the planet Venus across the disc of the sun, occurring about once every century

Venus tables: astronomical tables in the Dresden Codex describing the phases of Venus

Venus transit: same as Venus passage; an "eclipse" of the sun by Venus

vortexes: spiral organizing structures for spiritual energy

World Mountain: center of the earth in many ancient traditions

World Tree: perpendicular organizing structure for the creation of the cosmos

Xaman: Mayan name of the deity ruling the Northern direction

Xbalanque: one of the Hero Twins in the Popol Vuh, meaning "jaguar"

Xiuhtecuhtli: Aztec god of fire and time, ruler of the ninth day sign of the uinal and the first day of the trecena

Xochipilli: Aztec god of flowers, ruler of the eleventh day sign

Xochiquetzal: Aztec goddess of flowers, ruler of the twentieth day sign

Xochitl: Aztec name for the twentieth day sign, meaning "flower"

Xolotl: Aztec deity ruling the seventeenth day of the uinal, meaning "dog-headed monster," twin brother of Quetzalcoatl

Yaxkin: Mayan name for the center of the Four Directions

year bearer: day sign ruling the first day of a new year

Yggdrasil: Norse name for the World Tree, imagined as a huge ash

yin/yang: Chinese names for the dark/light, female/male polarity of the cosmos

yin/yang fields: projections of yin/yang dualities onto the earth's surface

Yohualticitl: Aztec goddess of birth, ruler of the eleventh day of the trecena

Yucatec: originating on the Yucatán peninsula

Yum Kax: Mayan name for the corn god

Zapotecs: ancient Mesoamerican people who discovered the tzolkin; they lived in the state of Oaxaca and built the ceremonial center of Monte Alban

zenith: high point, when the sun is directly above a certain location

zero: mathematical concept developed by the Maya around the time of Christ

Zoroaster: Persian religious leader who introduced a dualist yet monotheist religion

Bibliography

Argüelles, José. *The Mayan Factor: Path Beyond Technology.* Santa Fe: Bear and Co., 1987.

Argüelles, José. *Time and the Technosphere: The Law of Time in Human Affairs.* Rochester, Vt.: Bear and Co., 2002.

Balin, Peter. *The Flight of Feathered Serpent.* Wilmot, Wisc.: Arcana Publishing, 1978.

Bays, Brandon. *The Journey: A Road Map to the Soul.* New York: Pocket Books, 2001.

Brotherston, Gordon. *Book of the Fourth World: Reading the Native Americas Through Their Literature.* New York: Cambridge University Press, 1992.

Calleman, Carl Johan. *Maya-hypotesen, Svenskarnas roll för Gaias födelse år 2012.* Self-published in Swedish, 1994.

Calleman, Carl Johan. *The Theory of Everything: The Unification of Science Based on the Evolution of Consciousness.* Unpublished manuscript, 1997.

Calleman, Carl Johan. *Solving the Greatest Mystery of Our Time: The Mayan Calendar.* London and Coral Springs, Fla.: Garev, 2001.

Coe, Michael D. *Breaking the Maya Code.* London: Thames and Hudson, 1992.

Coe, Michael D. *The Maya.* London and New York: Thames and Hudson, 1993.

de Landa, Diego. *Yucatán Before and After the Conquest.* New York: Dover, 1978.

de Paz, Marco and Marcus. *Calendario maya: el camino infinito del tiempo.* Guatemala: Ediciones Gran Jaguar, 1991.

Freidel, David, Linda Schele, and Joy Parker. *Maya Cosmos: Three Thousand Years on the Shaman's Path.* New York: Morrow, 1993.

Hail, Raven. *The Cherokee Sacred Calendar: A Handbook of the Ancient Native American Tradition.* Rochester, Vt.: Destiny Books, 2000.

Jenkins, John Major. *Tzolkin: Visionary Perspectives and Calendar Studies.* Garberville, Calif.: Borderline Sciences, 1994.

Jenkins, John Major. *Maya Cosmogenesis 2012.* Santa Fe: Bear and Co., 1998.

Johnson, Kenneth. *Jaguar Wisdom: Mayan Calendar Magic.* St. Paul, Minn.: Llewellyn, 1997.

Maor, Eli. *June 8, 2004: Venus in Transit.* Princeton, N.J.: Princeton University Press, 2000.

Miller, Alice. *For Your Own Good: Hidden Cruelty in Child-Rearing and the Roots of Violence.* New York: Noonday Press, 1990.

Morton, Chris, and Ceri Louise Thomas. *The Mystery of the Crystal Skulls.* London: Thorsons, 1997.

Nisbett, Richard E. *The Geography of Thought.* New York: Free Press, 2003.

Popol Vuh: The Mayan Book of the Dawn of Life, trans. Dennis Tedlock. New York: Simon and Schuster, 1985.

Roys, Ralph. *The Book of Chilam Balam of Chumayel.* Norman, Okla.: University of Oklahoma Press, 1967.

Schele, Linda, and David Freidel. *A Forest of Kings: The Untold Story of the Ancient Maya.* New York: William Morrow, 1990.

Schele, Linda, and Mary Ellen Miller. *The Blood of Kings: Dynasty and Ritual in Maya Art.* New York: George Braziller, 1986.

Scofield, Bruce. *Day Signs: Native American Astrology from Ancient Mexico.* Amherst, Mass.: One Reed Publications, 1991.

Scofield, Bruce. *Signs of Time: An Introduction to Mesoamerican Astrology.* Amherst, Mass.: One Reed Publications, 1994.

Shearer, Tony. *Beneath the Moon and Under the Sun.* Santa Fe: Sun Books, 1987.

Thompson, J. Eric S. *A Commentary on the Dresden Codex.* Philadelphia: American Philosophical Society, 1972.

Waters, Frank. *Mexico Mystique: The Coming Sixth World of Consciousness.* Chicago: Swallow Press, 1975.

Internet Resources

Carl Johan Calleman's Internet address is www.calleman.com
(English/Swedish), where you may download a calculator for determining tzolkin days and tuniversaries.

A Web discussion group led by Sharon Jorgenson for those who have
read this book is available at
groups.yahoo.com/group/enlightened_future.

Information about the Oneness Celebration is available at
www.OnenessCelebration.com.

The Web page for Kalki (see chapter 9) is www.livinginjoy.com.

Index

Copán, 6

Cortés, Hernán, 2–3

cosmic information, 103–4

Cosmic Plan, 113, 173, 177,
    193–201. *See also* destiny

Cosmic Pyramid, 112, 163–64,
    185, 186–87

cosmology, 85–86, 129,
    133–36

Cozumel, 34

creation
    completion date, 144, 146
    divine plan and, 105
    thirteen-day count and, 20

creativity
    brain hemispheres and, 53–54
    Days and, 223, 225
    destiny and, 188–89
    influence of wave patterns, 27–29,
      49–50, 112

Creator God. *See* religions

Crete, 22, 44

crown chakra, 204

cult of Kukulcan. *See* Quetzalcoatl

Curry lines, 60, 202–3

Dark Ages, 39

Day Lords, 16, 237, 238–39

Days and Nights
    chart of, *130*
    creativity and, 223, 225
    frequency increase and, 110
    phases in development and, 123
    progression through Galactic
      Underworld, *145*, 146–58, *156*
    recessions and, 226

    in Sacred Calendar, 122
    Universal Underworld and, 217
    See also economies

day signs
    Aztec, 14, 15, 16, *78*, 121, 238–39,
      242–43, 250–51
    links to body parts, 98–99, *99*
    Mayan, 14, 15, 16, *78*, 121,
      238–39, *242–43*, 250–51
    practical uses of, 236–44

De Harmonice Mundi, 52

deities. *See* specific deities

delta waves, 57

democracy, 117, 149, 154, 261

Denmark, 44

de Paz, Marco and Marcus, 249,
    250–51

depressions, economic, 226, 231

Descartes, René, 47

destiny, 187–92, 219–20. *See also*
    divine plan

Deutero-Isaiah, 67

divination, 235

divine plan, 104–5, 168, 187–92.
    See also destiny

divine time, 111

dowsers and dowsing, 60

Dreamspell count, 126

Dresden Codex, 4, 127, 209

dualism, religions and, 7

dualist mind
    demise of, 217–18
    healing and, 182–83
    influence of Underworlds on, 115,
      165–66, 172
    problems created by, 196–97

Jainism, 80
Japan, 176
Jehovah's Witnesses, 156, 157–58
Jerusalem, 178–80, 262
Jesus Christ, 18, 68, 71–72, 84, 153
Jewish calendar, 112
Jews, 179
Johnson, Kenneth, 237
judgment, 185–86
Judgment Day, 191
Jung, Carl, 105

Kalki, 197–98
katuns, 245
Kepler, Johannes, 52
Khan, Genghis, 41–42
Kingdom of God, 68
King Menes, 22
Kukulcan, 7, 9, 74, 75, 77, 79. *See also* Quetzalcoatl
Kuwait, 256
Kuznets, Simon, 228

Lacandon groups, 9
In Lak'ech, 86–87, 186
Latvia, 42
left-brain hemisphere. *See* human brain
Long Count
    beginnings of, 17–21
    completion date, 144, 146
    depiction of, *18*
    emergence of higher civilizations and, 21–26
    prophecy and, 33
The Lord of the Rings, 151–52

love, 185–87
LSD, 184

magnetic pole shifts, 102, 160–61
Maharishi Mahesh Yogi, 170
Mahavira, 82
Maitreya, 194
Maldek, 102
Mammalian Underworld, 96, 114
material wealth, 189
mathematics, 11, 101
The Mayan Calendar, 249, 250–51
The Mayan Factor, 102
Mayans
    books of, 3, 6–7
    collapse of, 40
    creation and, 73–76, 89–90
    cultural context of, 2
    depiction of gods as tzolkin combinations, 124–25
    history of, 5–10
    legacy of, 1–16
        See also day signs; Sacred Calendar
Mayapan, 8
Medicine Wheel, 35, 49, 202
meditation, 167–70, 211–13, 248–49
Mediterranean, 82–83
men, 180
Menes, King, 22
Menorah, *73*, 73
Mesoamerica
    effect of the conquistadors, 2–3
    gods of, 20–21
    maps of, *3, 7*
    region defined, 2, 3

Ninth Underworld. *See* Universal Underworld

Nisbett, Richard, 61

Northern Hemisphere, 31, 147–48

numbers
  one, 83–86
  one hundred and eight, 199–200, 248
  role in Mesoamerican cosmology, 83–86
  seven, 46, 76, 161–62
  significance of, 98
  thirteen, 18–21, 19, 46, 83–86
  twelve, 83–86

Oceania, 69

octahedral structure, 59

odd-numbered heavens, 72–77

Olmecs, 77

Ometeotl/Omecinatl, 19, 21, 24, 42, 43, 45, 225

one, 83–86

One Giver of Measure and Movement. *See* Universal World Tree

one hundred and eight, 199–200, 248

One Intelligence, 63–64

Oneness Celebration, 210–14

One Source of Limits and Energy. *See* Universal World Tree

Osiris, 48

Ottoman Empire, 150

Oxlaj, Don Alejandro, 207–8

Palenque, 6, 8–9

patriotism, 141

Paul the Apostle, 38, 72

peace, 172–80

Pentagon, 148

Perls, Fritz, 184

personality traits, 240–44

Peru, 32

Pharaoh Djoser, 22

pilgrimages, 82

Pilgrims, 46, 174

pineal gland, 52

Planetary Round of Light, 112–19

Planetary Underworld
  burnout and, 111
  development of democracy, 154
  duration of, 96
  energy changes of, 93
  fantasy world of, 159–60
  Gulf War and, 255–56
  historical changes and, 106–9
  Industrial Revolution and, 226
  influence on Galactic Underworld, 139
  influence on modern minds, 165–66
  recessions and, 227
  ruling deities, 107
  unitary consciousness of, 117
  United States and, 175
  violence during, 147
  yin/yang polarity of, 114

Poland, 42

polarities. *See* yin/yang polarity

pole shifts, 102

Pope Gregory, 11

Popol Vuh, 7

postclassical period, 9

BOOKS OF RELATED INTEREST

THE MAYAN FACTOR
Path Beyond Technology
by José Argüelles

TIME AND THE TECHNOSPHERE
The Law of Time in Human Affairs
by José Argüelles

MAYA COSMOGENESIS 2012
The True Meaning of the Maya Calendar End-Date
by John Major Jenkins

GALACTIC ALIGNMENT
The Transformation of Consciousness According to
Mayan, Egyptian, and Vedic Traditions
by John Major Jenkins

THE MYSTERIES OF THE GREAT CROSS OF HENDAYE
Alchemy and the End of Time
by Jay Weidner and Vincent Bridges

CATACLYSM!
Compelling Evidence of a Cosmic Catastrophe in 9500 B.C.
by D. S. Allan and J. B. Delair

SECRETS OF MAYAN SCIENCE/RELIGION
by Hunbatz Men

RETURN OF THE CHILDREN OF LIGHT
Incan and Mayan Prophecies for a New World
by Judith Bluestone Polich

Inner Traditions • Bear & Company
P.O. Box 388
Rochester, VT 05767
1-800-246-8648
www.InnerTraditions.com

Or contact your local bookseller